Artificial Intelligence: Foundations, Theory, and Algorithms

Series editors
Barry O'Sullivan, Cork, Ireland
Michael Wooldridge, Oxford, United Kingdom

More information about this series at http://www.springer.com/series/13900

Paula Boddington

Towards a Code of Ethics
for Artificial Intelligence

Springer

Paula Boddington
Dept of Computer Science
University of Oxford
Oxford, United Kingdom

ISSN 2365-3051   ISSN 2365-306X   (electronic)
Artificial Intelligence: Foundations, Theory, and Algorithms
ISBN 978-3-319-86905-6   ISBN 978-3-319-60648-4   (eBook)
DOI 10.1007/978-3-319-60648-4

Printed on acid-free paper

This Springer imprint is published by Springer Nature
The registered company is Springer International Publishing AG
The registered company address is: Gewerbestrasse 11, 6330 Cham, Switzerland

This work is dedicated to the memory of my dear friend and colleague, Professor Pamela Sue Anderson, 1955–2017

Foreword

Few academic disciplines have suffered as many reversals of fortune as artificial intelligence (AI). Early AI researchers seemed to make rapid progress and became convinced that they were on the fast track to the grand dreams of AI—only to find that progress petered out, amid claims that AI was nothing more than modern-day alchemy. Successive waves of subsequent AI technologies have promised progress, but in the end progress proved possible only on very narrow problems.

We are currently in one of the periodic boom times for AI. There has been genuinely impressive progress in the area of machine learning, prompted in part by the availability of cheap computer power and big data and in part by scientific breakthroughs. This progress has caused massive media hype, and there has been much speculation about the possibility that the AI problem has finally been solved and that we will now see swift progress to the grand dream of AI. And this possibility has in turn caused some well-respected pundits to publicly voice concerns about the dangers that rapid progress to AI might bring.

Whether or not we are on the fast track to intelligent machines, it seems that recent advances in AI bring with them real challenges. There will very likely be important societal challenges arising in areas such as (un)employment, privacy, healthcare and autonomous weaponry. Scientists, technologists and policy-makers need to be aware of these issues and must be able to respond to them. Ethical considerations naturally come to the fore—which brings us to this volume.

Ethical codes of practice have a long history in science but not an untroubled one. If we want to think about ethics for AI—and I believe we should—then we should do so informed by the historical experience of other disciplines, and with a realistic understanding of the issues that ethical codes of practice themselves raise. I am delighted to have the opportunity to introduce this volume, written by a philosopher with a long track record of research in medical ethics. The book explores in detail these issues, and should be read by everyone with an interest in the future of AI.

Oxford, UK   Michael Wooldridge
June 2017

Foreword

It is a privilege and a pleasure to introduce Paula Boddington's new book, which provides a wonderful new resource for anyone interested in ethics or in artificial intelligence (AI), and is surely essential reading for anyone interested in their overlap. Although its title might suggest a relatively narrow and professional focus on the development of ethical codes for AI research, it is very engagingly written, and full of valuable insights into ethics more generally: a book that deserves a wide readership, in school and university teaching as well as amongst relevant professionals.

Ethics has ancient roots, reflecting its perennial and central role in human life, while AI has developed significantly only in the last half-century. And it is the explosive growth of the power and potential of AI over the last decade or so that has brought to public prominence the crucial importance of addressing the many ethical issues that it raises. Within a very few years, we could be living in a world in which a huge proportion of the decisions that affect our lives—from financial markets to transportation and from healthcare to military operations—are either made or significantly informed by AI systems.

Understandably, therefore, a host of projects have recently sprung up around the world to try to anticipate and ameliorate the many ethical, social and legal problems that AI could spawn. But much of what has been said on these topics, including most of the output that has been the focus of media 'hype', has been philosophically simplistic and crude. There is, after all, no reason to expect that high-profile entrepreneurs or pioneers of AI—even if their words command widespread public attention—will have special insights into complex ethical issues and the many ways in which these interact with personal, cultural and legal perspectives. We do not expect arms manufacturers to be authorities on the ethics of war, or surgeons or geneticists to be authorities on the ethics of medicine. Indeed, our doubts are likely to be all the greater when these people may have economic or other motives for their views, or are futuristic enthusiasts whose interests are highly untypical.

When considering the ethics of future AI, moreover, it is even less appropriate to base our judgements primarily on the views of company bosses or technical researchers, because we are imagining a future in which the bounds of machine intelligence press further in many directions, often going well beyond what anybody can currently foresee. Thus, our understanding of the problems we will face needs to be moulded far more by our ethical and philosophical landscape than by current technological details. And hence to assist these debates, there is a crucial need for resources that can help participants to learn the lessons that have been garnered down the years by ethicists and philosophers, working in areas that share some of the same characteristics, and to examine carefully and critically how these lessons can impact on AI and the ethical dilemmas that it is likely to provoke. This new book provides just such a resource, bringing rich insights from a moral philosopher who has many years of experience working at the cutting edge of medical ethics, probably the closest relative of AI ethics in terms of the novelty of the questions it raises, its capacity to impact on our view of ourselves and on many aspects of our lives and its perspective towards an unknown and constantly evolving future.

Very importantly, Boddington is not writing from a particular narrowly defined ethical perspective and is not attempting to promote a specific theory about AI ethics. Indeed, the book exhibits a profound awareness of the complications and pitfalls of ethics and the dangers of hubristic claims to comprehensive answers or simplistic solutions. It is also written in a delightful style, with humour, verve and a willingness to puncture theoretical balloons with down-to-earth observations. As a philosopher myself, I can see clear evidence here of many years' teaching of ethics, as common errors are highlighted, cautionary points judiciously made and pompous vacuities (which are all too common in public ethical pronouncements) exposed.

Quite apart from the focus on AI, the early chapters provide an excellent introduction to ethical thinking, from which many other philosophers—both teachers and students—could learn much. One particularly nice feature of the book is the many boxes containing succinct self-contained discussions of specific issues, including algorithmic bias, transparency, moral disgust, relativism, health and social change, narrative in policy-making, ethical arms races, administrative evil and virtue ethics. These quickly enable the reader to absorb a large number of insightful points over a wide range and also help to make the book a real pleasure to dip into. Another nice feature is the frequent raising of the question '*What's this got to do with AI?*', spelling out explicitly both how the various ethical issues are relevant to AI and how distinctive developments in AI can bring new factors into older ethical discussions.

Later in the book, the discussion of codes of professional ethics is well informed and judicious, drawing valuable lessons from past experience of such codes, citing evidence from psychology and other relevant fields and highlighting potential problems that are commonly overlooked. The development and application of such codes is by no means straightforward, and they can bring dangers as well as benefits. As applied to AI, a number of distinctive problems emerge, ranging from the distributed and unprofessionalised nature of AI research, the concentration and imbalance of resources, possible mixed motivations of the various actors and the enormous range—both anticipated and unforeseen—of potential applications. All of these make it hard to extrapolate from the past development of ethical codes in, for example, civil engineering or medicine.

An entire chapter is devoted to another distinctive problem, namely, the development of a code of ethics for a technology that is rapidly evolving—both technologically and in terms of its social and economic impact. Profound social or economic change characteristically leads to moral uncertainties and disagreements between groups within society. Working out codes of ethics that can guide us through these minefields, in a way that can command general assent, poses a serious problem. Issues of values and responsibility are difficult enough with a technology that crosses all international and cultural boundaries, but they are made even more intractable in the context of ubiquitous change. Boddington plausibly suggests that diversity of participation in these developments at least provides an important part of any solution. She then goes on to provide a helpful guide to 'characteristic pitfalls in considering the ethics of AI', again making it easy for readers to acquire valuable insights from her extensive thought and experience.

The book ends with a discussion of 'suggestions for how to proceed' in developing codes of AI ethics, emphasising factors such as proper institutional backing, public trust, diversity in participation, transparency, communication and appropriate procedures for revision and critique. Again Boddington brings to bear her experience as a moral philosopher, carefully and helpfully explaining relevant considerations. Overall, the book makes abundantly clear that there are no easy answers, and indeed, we are left only too aware of the many different questions that need to be faced if we are to stand any chance of developing codes of ethics that might significantly ameliorate the risks arising from the future development of AI. But I have no doubt that our chances will be greatly increased if those concerned pay serious attention to this excellent book.

Oxford, UK   Peter Millican
June 2017

Preface

This work has been funded by the Future of Life Institute, as part of the project Towards a Code of Ethics for Artificial Intelligence Research, awarded under the FLI's 2015 project grants for Beneficial AI, from funds donated by Elon Musk and the Open Philanthropy Project. Without the Future of Life Institute's generous support, this work would not have been possible.

The project has been hosted in the University of Oxford's Department of Computer Science, where I have been working with Professor Mike Wooldridge and with Professor Peter Millican of the University's Philosophy Faculty. The work, therefore, was only possible with the generous support of the Future of Life Institute and with the support and input from Mike and Peter.

Another important source of inspiration and ideas has come from some involvement with the Institute of Electrical and Electronics Engineers (IEEE) Global Initiative for Ethical Considerations in Artificial Intelligence and Autonomous Systems. Even my small participation in this programme has helped me considerably in my quest to think through the challenges and opportunities for trying to ensure that developments in AI are ethical and beneficial.

Part of my involvement with the IEEE was as a member of a group attempting to address issues of participation in their work. In attempting to ensure that a fair and balanced view of issues is obtained, there are important and obvious reasons in any field, and in ethics especially, that no voices are excluded from discussion, and that one approaches issues with an open and critical mind. This is an ideal but rarely a reality. Hence, this book certainly has multiple shortcomings and blind spots. Its length guarantees that it certainly misses many important issues, and there are many issues which it has only been possible to indicate in outline.

However, there are worldwide conversations and initiatives regarding important ethical issues in artificial intelligence (AI) happening right now, and it seems important to contribute in a timely manner to such debates. It is in this hope that this book has been written. Indeed, if omissions and flaws are apparent, spotting them can only be a good thing, if this contributes to further advancement in this field.

As with any work, of course, I came to these issues with a particular background and concerns. With a long-standing study of ethics, both theoretical and applied, I came to the opinion that some further elucidation and explanation of some basic philosophical and foundational issues in ethics might assist; I frequently noticed that there were points of disagreement and debate in the ethics of AI which hinged on such issues. Notwithstanding the ever-present tendency to hype up the issues of AI, whether this concerns its technical possibilities, or its ethical and societal impacts both positive and negative, I also came to the view that, yes, AI presents us with ethical questions but more than this. Some of the questions that AI presents us with are not only of the form 'should we do this or should we do that?'; in many respects, the challenge of AI relates to some of the central presuppositions about the nature of human agents and our place in the world, presuppositions that lie at the heart of ethics itself.

I first used computers when working as a laboratory technician in the physics research department of a photographic company, punching data into bits of cardboard and feeding them into machines the size of fridge freezers. Since then, much of my work has concerned ethical issues in various fields, including medicine, genetics and genomics. I learned much from working in interdisciplinary teams and research centres addressing ethical, legal and policy issues, with sociologists, psychologists, discourse analysts, lawyers, research scientists and others, and, therefore, owe a debt of gratitude to what I learned from my many colleagues from years past. The work of this book has therefore been influenced by noting some parallels with these fields, which concern the rapid development of technology, science, and its applications, where questions are raised, at some deep level, about how we think of ourselves as human beings. The implications for AI are I hope obvious. A chief task is to understand how issues in AI and in the technological, societal and cultural changes associated with it are related to other issues and to our world, which is ever changing, yet in other ways, ever the same.

We are focusing on the challenges of developing codes of ethics for AI researchers. This naturally involves understanding in broad terms what the ethical questions are, in different fields of AI and for AI more generally. But it also involves considering what the purpose of codes or regulations might be and how to produce workable and effective codes or regulations. This involves looking not simply at the content of such codes but at questions such as who's involved in their production and critique and who's involved in their application. It involves considering also the surrounding social, economic, cultural, legal and political conditions which form a backdrop to the development of AI and of any codes of ethics concerning AI.

Many thanks to those who have made this work possible.

Oxford, UK   Paula Boddington
May 2017

Contents

List of Abbreviations

AI	Artificial Intelligence
AoIR	Association of Internet Researchers
IEEE	Institute of Electrical and Electronics Engineers
IIIM	Icelandic Institute for Intelligent Machines
UBI	Universal Basic Income

Chapter 1
Introduction: Artificial Intelligence and Ethics

Abstract This brief introduction sets a context for the subject of the book: the challenges that arise in developing codes of ethics for artificial intelligence (AI). To start with, an overview of some of the concerns about AI and current developments in AI. AI encompasses a wide range of applications, of very varying natures, which means that there will be complex debates about its benefits and risks. Some of the many and varied current initiatives concerned more specifically with AI and ethics are introduced briefly. There are diverse approaches to tackling ethical issues in AI which may complement the development of codes of ethics, and these too are briefly outlined.

1.1 Why Ethics in AI? Why Now?

This book is concerned with the question of how to develop codes of ethics in artificial intelligence. But why look at this now?

Hardly a day goes by when there is not a prominent media story about the risks and benefits of artificial intelligence. Alongside accounts of the technological promise of AI are warnings about its perils. Even those who herald a glorious future with AI often paint a picture of profound social and individual change. But should we be worried? And if so, what should be we worried about in particular? And precisely how worried should we be?

For one obvious feature of AI is its broad nature and manifold applications. There is also some problem even in defining it, for it's often been observed that a technological capacity is heralded as AI until it's in place, then, as John McCarthy, the computer scientist who coined the term 'artificial intelligence', put it, 'as soon as it works, no one calls it AI anymore' (Vardi 2012). The borderline between what counts as AI proper and other forms of technology can be blurred. Some AI systems are so embedded within technology that they are assumed and almost invisible. This also means that it's hard or impossible in many cases to say which ethical and other value issues are presented by AI per se, and which by other features of technology.

AI is already with us; many applications are fast developing and will be with us in the near to medium term. It is a disputed question when other even more advanced forms of AI including superintelligence will be here, if ever. Many consider that the emergence of superintelligence is an inevitability; there are the usual disputes about how long it will be before it is with us, as well as disputes about whether we should welcome its arrival, and if so, why. Philosophical and technical disputes converge, in the debates about whether we will ever develop AI which has consciousness and which is sufficiently complex, and in the right ways, to merit our moral concerns and protection.

AI may involve robotics of varying complexity; it may involve the manipulation of vast amounts of data; it may involve machine learning. It may involve how we relate to others; it may involve how as individuals we think, remember and reason. It may have implications for the organisation of the labour market; it may involve how we trade; who has access to information; and how. It has implications for the economy, for politics, for culture. It has implications for those who directly use AI, such as those who use a robot butler; it has implications for those more remote, as for example when an algorithm developed by machine learning helps makes public policy decisions, or unemployment attributable to the use of AI means a worker can no longer afford to pay their child's music teacher.

Artificial intelligence may be applied in many very different areas, and in different ways within the same area. In medicine, AI may be involved in computerised diagnosis of individual patients, or in algorithms to analyse vast amounts of data from thousands or millions of patients to understand the nature of disease and health. It may be involved in patient consultations and even therapy sessions with online or robotic responses. Robotic assistance with surgery can involve AI, for complex and delicate operations. It may be involved in remote monitoring of health, in mobile technology that gives patients information about their own conditions. It may be involved in nursing and care, with robotic assistants or companions. Robots are being used to assist people with autism to develop social skills. Robotic pets are being developed to provide companionship and mental stimulation to patients with dementia. Robotic limbs are being developed, as well as devices to enable patients with locked in syndrome and other similar conditions to communicate.

Machinery increasingly involves AI. Autopilots include elements of AI, and autonomous vehicles are imminently set for widespread use. The 'internet of things' connects household gadgets and other items using AI. Commerce involves AI, from automated trading agents in the stock market, to algorithms which tailor online advertising or sort the prices of airline tickets, both for buyers and for sellers. The use of robotics in manufacturing is long established, and capabilities are increasing. AI is moving into performing work that previously required not just manual skills but intellectual skills, as in legal research and accounting. Teaching also faces inroads from AI. AI may even be used in creative endeavours, such in writing literature; it's already being used in the composition of music.

There will hence be many different and complex debates to be had about the perils and benefits of AI and its applications. How then, should we think about developing any codes of ethics for AI?

1.2   Current Initiatives in AI and Ethics

There are so many initiatives currently underway concerning the ethical, social and legal aspects of AI that it would be a project in itself to list them all. Here I simply indicate examples of activities from across diverse categories.

There are of course academics from different disciplinary approaches working across many universities, including dedicated centres, such as the Leverhulme Centre for the Future of Intelligence at Cambridge (http://lcfi.ac.uk/), and the One Hundred Year Study of AI (AI100) at Stanford University (https://ai100.stanford.edu/). Such centres may get funding from a variety of sources.

As well as projects explicitly examining ethical issues in AI, there are also AI projects which themselves incorporate ethical objectives. These projects show an awareness of the potential problems of AI. For example, OpenAI aims to produce open source AI code under the belief that this is the best way forward to combat possibly malicious use of AI (http://open.ai/). The Machine Intelligence Research Institute (MIRI) states its aim as 'aligning advanced AI with human interests' (https://intelligence.org/).

Large and smaller corporations also have initiatives, such as the Partnership on Artificial Intelligence to Benefit People and Society, with collaboration from Amazon, DeepMind, Facebook, Google, IBM, and Microsoft (https://www.partnershiponai.org/); and the non-profit initiative AI Austin with collaboration between university, City Council and business (https://www.ai-austin.org/). There's also work by individuals in the professions, such as research examining bias in AI recruitment and the development of apps to investigate bias in algorithms (Clark 2016).

Work by government agencies is also underway, such as the White House Report on the Future of Artificial Intelligence (Executive Office of the President 2016) and a draft report on robotics and law by the Committee on Legal Affairs of the European Union (Directorate-General for Internal Policies 2016). Such governmental work tends to focus on wider societal issues such as employment, funding and economics, which naturally have ethical implications.

There is work by professional bodies such as the IEEE's Standards Association Global Initiative for Ethical Considerations in the Design of Autonomous Systems, which has a large number of strands of investigation motivated by their byline, 'Values By Design.' This is an ongoing project to produce both industry standards and discussion documents on various topics (http://standards.ieee.org/develop/indconn/ec/autonomous_systems.html). There's work by research funding councils such as the Engineering and Physical Science Research Council (EPSRC)'s Principles of Robotics which was produced in 2011 and designed to stimulate discussion (Boden et al. 2011).

There is also work by special interest and pressure groups from the Campaign to Stop Killer Robots, whose remit is obvious from the name (https://www.stopkillerrobots.org/), to the #HellNoBarbie campaign fighting against Hello

Barbie and other children's toys which can transmit and analyse children's conversations remotely (Taylor and Michael 2016).

There are projects funded by non-profit organisations, such as our project, which is funded, along with 34 other projects, with an AI grant awarded by the Future of Life Institute. The Future of Life Institute held a 5 day workshop in January 2017 at Asilomar in California, during which time they drew up a set of Asilomar AI Principles. These will be discussed in greater length in Chap. 8.

1.3   Codes of Ethics in Context: Other Approaches to Ethical Questions in AI

Even if codes of ethics for AI can be produced which are robust and effective, there are other strategies that are followed in the pursuit of ethical and beneficial AI. Indeed, there's an intimate relationship between different strategies, since a code of ethics might include injunctions or recommendations to pursue certain technological pathways, and because the technical realisation of ethical concerns is a *sine qua non* of achieving beneficial AI. Again, here I can only give a very brief overview.

1.3.1   Epistemic Strategies: Precision and the Reduction of Uncertainty

Work pursuing epistemic strategies includes moves to reduce levels of uncertainty regarding future development of AI, to render more precise strategies for dealing with AI possible. These may involve broad future-watching developments. Some researchers are applying methodology to track the speed of development of superintelligence or supercomputers, and to understand the risks and impacts of AI, such as the AI Impacts project run by Katja Grace (Grace).

A common strategy is to share and publish research results online, and to make results as broadly accessible as possible. Strategies to make the operation of AI transparent may also help to reduce uncertainty. There is also ongoing work that aims to clarify the associated ethical, legal and conceptual issues involved in AI to gain a clearer account of what problems must be tackled. One question to be addressed is how, given uncertainty about the development of AI, do we trade off considering questions about distant and uncertain developments, as against more immediate issues? Concern that discussion proceeds at a good pace is one reason why I am writing this book now, even though it's bound to be incomplete.

1.3.2 Technological Strategies to Ensure Safe and Beneficial AI

General strategies to make AI error free or safer are important elements in tackling ethical issues. Verification (ascertaining that an end product meets design specifications) and validation (ascertaining that an end product meets user needs and that the specifications were adequate for the environment of use) have obvious ethical implications. There are also various strategies to align AI with human values. Another common broad strategy is to try to ensure ultimate human control of AI. But whether this will be even possible for advanced forms of AI is subject to debate. Working towards trust between humans and AI is a variant of ensuring control.

1.3.3 Moral Strategies in the Pursuit of Beneficial AI

Some commentators have tried to mollify fears about AI. These include claims that AI presents no new ethical problems, so that there is nothing particular to fear. Or, moral threats of AI may be balanced against benefits, as with claims that AI might help us to make moral decisions better. Whether or not this is possible, and how, is a moot point. It's worth pointing out that if we need AI to help us make moral decisions better, this casts doubt on the attempts to ensure humans always retain control of AI. The gulf between these opposed approaches is perhaps typical of the difficult roller coaster terrain of AI ethics.

Other attempts at mollifying the threat of AI run up against the difficulty that one person's comfort is another person's panic. For example, claims that in the future we will be part human/part machine and live lives of leisure enjoying art and literature produced by machines may entice some but leave probably at least equal numbers running in panic for a commune in the hills far from any wifi signal, to eat berries, whittle wood and tell stories of the old days around the campfire. There are those who warn as loudly as possible about the moral dangers we are in from AI. Producing codes of ethics or sets of principles for the beneficial and ethical development and use of AI needs to be done against the backdrop of a realistic assessment of the issues and the appropriate level of moral concern.

But before we can proceed, we need to do some ground clearing. For when we say that AI presents us with ethical questions, what do we mean by 'ethics'? The next chapter will briefly set out some questions about ethics that we need to understand for our discussion. Following that, we'll turn to consider if AI presents any distinctive ethical challenges, before then turning to examine professional ethics in particular. We'll then be in a better position to consider what challenges lie ahead for developing professional codes of ethics for AI.

Chapter 2
What Do We Need to Understand About Ethics?

Abstract Consideration of ethical questions in AI requires an understanding of some central questions and ideas in ethics. This chapter provides an introduction to ethics which will be used as a basis for further explanation of the particular questions about ethics in AI. Ethics is sometimes seen entirely negatively as restricting developments, but can also be used more positively as assisting in the promotion of beneficial activities. Standard normative ethical theories are outlined, but the focus here is on spelling out underlying questions in ethics. We need to understand that there are diverse accounts of the root need for ethics, questions about the nature of ethical concerns, and questions about who, or what, is the proper object of our moral concern, all of which need to be addressed in thinking about AI. There are also contentious questions about the nature of argument and justification in ethics, including questions about moral relativism, which are especially pertinent to the issue of developing codes of ethics, and which we will need to consider carefully. The issue of transparency in ethics parallels concerns with transparency in AI. Questions about the nature of moral agency and moral motivation are also of prime relevance to discussions of AI.

From reading much of the literature on AI and ethics, and from taking part in many hours of discussions with a range of people from a variety of disciplinary backgrounds, I've realised more and more that there are some questions and issues in ethics which are omnipresent in many of these discussions, but which are not always articulated.

It is a central contention of this book that developments in AI require that we consider and perhaps reconsider some fundamental questions in ethics.

It's obviously impossible to present a full characterisation of ethics here. There is disagreement among philosophers on every one of the issues we will discuss. The points raised are pertinent to codes of ethics for AI; to considering some of the central ethical questions of AI more generally; as well to as the thorny question of whether or not, and how, we can build ethics into machine behaviour.

© Springer International Publishing AG 2017
P. Boddington, *Towards a Code of Ethics for Artificial Intelligence*,
Artificial Intelligence: Foundations, Theory, and Algorithms,
DOI 10.1007/978-3-319-60648-4_2

2.1   A Preliminary Plea: Ethics Is Not About 'Banning' Things

Very often, talk of 'ethics' and in particular 'ethical regulation', conjures up the idea that 'ethics' is simply out to stop activity, prohibit or mandate various actions. In some circles, the word 'ethics' has attained negative connotations (Bowie 2009). Indeed, some 'ethical' regulation can with some justification be found guilty of excessively hampering valuable research—and to this extent then, 'unethical' (Atkinson 2009). We'll directly consider later the possible negative impacts of codes of ethics. But this 'spoilsport' notion of ethics is limited. Ethics can and should be seen more positively as helping to promote or enhance an activity.

Note that we may recognise that an activity merits our attention and requires ethical discussion, without deciding in advance that this means it's going to turn out to be problematic. We need to be aware of how changes are impacting on our values. Self-awareness, both as individuals and as societies, is itself of value. In considering ethics in the context of artificial intelligence, amidst talk of the possibility or otherwise of self-aware machines, we must here of all places recognise its value.

2.2   Normative Ethical Theories

Many accounts of practical ethical questions will start off with a broad characterisation of different normative ethical theories. These are accounts of how to act; in other words, theories about the basis for making decisions in ethics. The three most commonly outlined theories are:

Consequentialist theories, which broadly claim that the right action is the one that brings about the best consequences. This is most commonly held as some form of utilitarianism, which aims to bring about the greatest balance of happiness over unhappiness, or pleasure over pain, for the largest number of people.

Deontological theories, which claim that what matters is whether an action is of the right kind, that is, whether it is in accordance with some general overarching principle, or with a set of principles, such as 'do not take innocent life', 'do not lie', and so on.

Virtue ethics, which focuses of the character of the ideal moral agent, and describes the range of different virtues such an agent has, and, broadly, claims that the right thing to do in any given situation is to do what the fully virtuous person would do.

There is much that can be said about these theories, their differing interpretations, and the vexed question of how to 'apply' theory to practice in ethics. However, normative ethical theories will not be our focus in this book. Important elements of morality which lie behind and outside these theories need to be examined to gain a fuller appreciation of the ethical challenges of AI.

2.3   Ethics and Empirical Evidence

Ethics deals with normative issues; it is not purely descriptive of empirical reality. Normative issues are ones we feel have a certain weight and import, although it's surprisingly hard to characterise precisely what the weight and import of ethical issues are, and there is philosophical disagreement about whether ethical issues should always override other considerations.

The normative nature of ethics means that simply describing the way people act will not give an account of ethical action. Ethics requires discrimination between ways of acting and of being. Nonetheless, empirical questions about how we do think and act, and the possibilities of human psychology and society may be relevant to any consideration of ethics, for a variety of reasons. As the philosopher Kant observed, 'Ought implies can'—we can't require an individual person, humans in general, or indeed, machines, to do something that they *cannot* do (Kant 1998). We need to know what's possible for human action, what might be effective strategies for assisting with obstacles to moral judgement and action, what effects there might be on human health and wellbeing of various possible policies, what pitfalls of action and judgement await us as we strive to think and act for the good, and so on.

What this means for AI: We need to think carefully about what relevant empirical evidence we have to collect to assess the impact of AI. This is harder than it might seem, and for interesting reasons. The evidence we need to consider is about the impact upon complex, feeling, living beings, immersed in sophisticated, dynamic cultures; it's about human beings who only partly understand themselves, and who only partly understand their own cultures and societies. It's about untangling what appears to be the case, and what is the case. AI, now and in the future, is deeply embedded within other technologies and with social practices; so measuring impact and attributing it to AI will be extremely challenging.

2.4   So Why Do We Even Need Ethics?

It's worth pondering this, for there are different answers. Often these answers are strongly shaped by the disciplinary background of the questioner, be it sociology, anthropology, evolutionary biology, philosophy. Again, the aim here is not to produce 'an answer', but to indicate that whatever answer is given, it will reveal issues of central relevance to questions of ethics and AI.

One broad brush answer is that ethics exists because the world is not perfect, and we think we could improve it if we tried hard enough. But if this were the only ethical problem, then we'd simply need to sort out how to improve the world, and then, improve it. Simple! There are at least two further problems.

One, the world is imperfect *in a really complicated way*. It's often hard to work out what *precisely* is wrong, let alone have a clear idea of what to do about it.

And two, *we* are not perfect, whether as individuals, or as groups. Even when we know what do to, we don't always do it—there is a problem with moral motivation. We lament with Rodney King, 'Why can't we all just get along?' And note, we ourselves often think we personally could have done better. We have some idea of what St Augustine meant when he prayed, "Grant me chastity and continence, but not yet." (Augustine 2014)

Many philosophers have considered that we need morality because things are 'inherently such that things are liable to go very badly' (Warnock 1971), and that we can't sort this out if left entirely to our own individual whim. Morality is a 'device for counteracting limited sympathies' (Mackie 1977). But even among those thinkers broadly in this tradition of 'double deficit' where both we and the world are broken, there are significant disagreements. For instance, not all agree on what particular shortcomings we have as humans. Some focus on problems with reasoning, some on our emotional responses. Some consider that if only we fix bias, we'll do the right thing. Some consider the end result of fixing bias must be some kind of equity. Some are idealistic utopians about human perfectibility. Some consider that the price of civilisation will always be a certain amount of discontent (Wiseman 2016; Freud 2002). And so on.

Morality as a Solution to Competition for Scarce Resources: What's AI Got to Do With It?

It would make a rather interesting project of its own to consider how different models of the function of morality, and the concomitant picture of human nature and the world, interacted with the development of AI.

But to illustrate, and to see how deep questions about AI and ethics go:

Suppose you consider that we need morality to combat our bias towards ourselves and our kin, given that there is scare competition for resources in the world and these need to be shared with some measure of fairness. Then, we usher in a glorious future of advanced AI.

We'd still be biased, of course. So do we outsource our ethical judgements to AI? Note the precise details of how we do this will depend not just on how we understand our own biases, but also on how we understand the ultimate goals of morality.

And even if we do this, why would we obey the AI, given our shortcomings in moral motivation? So, should we tie ourselves in to being forced to obey the AI? Should we go for individual enhancement via AI to combat this bias, so each of us is morally 'corrected'? In which case, we no longer have the same picture of the need for morality.

And what is the point of AI if it can't solve the problem of scare resources? So now we live in abundance. But abundance of what? Material goods, perhaps; but what do we do all day? Many scenarios foretell mass unemployment; goods aplenty, jobs scarce. If morality combats a problem of resource

(continued)

distribution, and the resources which are scarce change, we might need quite different moral tricks up our sleeves to address the rather different challenges of plenty.

Another case: suppose we say the task of morality is to make sure that each person lives a decent life, despite scarcity in the world and human shortcomings.

Or suppose we say the task of morality is to make sure that there is as little suffering in the world as possible, despite scarcity in the world and human shortcomings.

The former implies that AI should be geared towards making sure that those rendered unemployed by machinery all have good lives, suited to their individual situations. It might even double back on the use of AI to prevent individual misery.

The latter leaves it wide open that AI might be geared towards trying to ease out of existence the class of people who don't cope well with AI induced redundancy, whether by a programme of eugenics or of enhancement.

Note that an account of ethics will explicitly or implicitly rest upon underlying views of moral agents—us—and of our place in the world. It will implicitly rest upon underlying views of the value and nature of that world. It will implicitly rest on views of the relationship between us, as moral agents, and other moral agents, and the rest of the world. It will rest on an account of what inclines humans to behave badly, and what enables them to behave well. It will rest upon assumptions about how good a job we can do of perfecting 'human nature' and the world. Such underlying issues will surface, at some point, in discussions of codes of ethics for AI. They may be in disguise. But they will be there.

What this means for AI: The take home message is that understanding ethics means understanding moral agency. And how we understand human agency in particular, and agency in general, is a critical question in AI.

2.5 So, With What Sort of Issues Is Ethics Concerned?

Let's start with a popular answer to this. Ethics concerns important questions of welfare and harm, or if you prefer, pain and happiness, along with important questions of justice and fairness. The questions of justice and fairness bring with them questions about balancing the interests of individuals and groups. These questions tend to predominate many formal academic discussions of ethics, but there are other values which are important to recognise, such as the value of loyalty, of (justified) respect for authority, and ideas that relate to some notion of sanctity or purity—drawing what are seen as proper boundaries between different elements of our world (Haidt 2013).

It's easy to get an intuitive handle on many of the core ethical values, but very hard to specify them in detail without running against problems. Let's take an example: human health. This seems like a sound moral goal to pursue. But it turns out to be impossible to characterise health without addressing many other value issues. Should we have as a goal of human health, the maximal extension of human life, the postponement of death for as long as possible? Yet some would consider that there is a 'natural' termination to human life, others not. Should we extend the life of someone with such advanced dementia that their personality is no longer apparent? Addressing such a question involves asking and answering questions about the nature of the human person over time, and questions about how one person relates to their past and future selves, and to other people. These questions then rest upon accounts of human nature, human agency, what it is we value about life, and about personhood. And such questions come to the fore in many questions involving the development and application of AI.

Likewise, consider the fundamental question of whether we should aim at maximising happiness, in the sense of maximising pleasure, in humans. On many views, this gives an impoverished account of what human life should be about. Surely we want to do more than sit around with the pleasure centres of our brains firing away? Or is this really what we do actually want? Do we then need to address questions of the meaning of human life? Of its point?

There's also the question of where the boundaries lie between questions of ethical value, and other sorts of value, such as aesthetic value, and political questions. In drilling down to fine detail, there will be substantial questions raised. For example, how does the value of equality play out in relation to the complex and heated debates about what behaviour does and does not count as 'sexist'? Translation of values into the behaviour of AI has already raised many detailed questions of interpretation, such as the question of how Siri responds to sexist 'banter' (Fessler 2017).

What this means for AI: These questions turn out to be utterly crucial in considering the replacement of human activity, whether in whole or part, by machines. To that extent, these are questions already raised by mechanisation, but the developments of AI heighten our concerns. We will discuss these issues later.

2.6   Who (or What) Is The Proper Object of Moral Concerns, and How Widely Should Our Concerns Extend?

It's easy to assume that ethics must have universal reach (however this is defined), and that a sound ethic has to reach beyond individual, tribal or group concerns. It's commonly held that everyone shares a universal ethic, but this is demonstrably false. The views of Aristotle are particularly influential among many moral philosophers currently, but he did not take a universal view of ethics, distinguishing not just between men and women, free man and slave, Athenian and barbarian, but also held that one had significant duties to one's parents and one's children, yet no particular duties towards grandchildren (Aristotle 1999). Many actual systems of ethics have different rules of behaviour for different classes of people.

Moreover, even for those who hold that moral demands apply universally, there's the question of who counts morally: humans as a species; or persons, a class which may include some who are not human and exclude some who are; or any creature that is capable of suffering; or wider still, as some environmental philosophers argue? The philosopher Immanuel Kant held that moral concern should extend equally to all rational beings, and that would apply to rational creatures from other planets. He might or might not then have added that it could apply perhaps to some forms of AI (Kant 1972).

What this means for AI: Could we ever have moral obligations to sophisticated artificial intelligence? This depends on the basis for our moral obligation, and for why others—other creatures, other machines—have value. On some views of human and the human brain, we are pretty much like calculating machines, like computers, with various goals built in. On such a view, it's then more feasible that we might build AI which, like us, has moral standing, and can act as a moral agent. But others hotly dispute the initial premise that this is a good view of what humans are like. These are not questions extraneous to ethics. These are questions which underpin any account of ethics we might have. Hence, the question of who merits our concern, has large ramifications for considering ethics in AI.

2.7 Four Domains of Ethics: Self, Friend, Stranger, World

For some, ethics is essentially about how we treat other people. The view that it's all about combating disparate interests under a condition of scarcity suggests this. Such views may posit an egoistic motivation, and often assume self-interest: we can act in any way we like, so long as we don't harm others, and it's assumed either that we always act in our own interests, or that it's none of anyone else's business, and of no moral import, if we don't.

But on other views, we may have ethical responsibilities towards ourselves. If you're someone to whom this seems counter-intuitive, simply ask if it's okay voluntarily to get rigged up to a machine that stimulates your pleasure centres, rather than actually acting in the world. This seems abhorrent to many people and a travesty of a good life; it may even seem a morally wrong waste of a life. Others of course, beg to disagree. This debate can be very polarised; I've noticed that those few undergraduates who tough it out and insist that they'd be rigged up to the pleasure machine often find others respond with horror.

What this means for AI: Note that the potential of AI to powerfully transform our sources of choice, value and pleasure raises issues which come very close to these concerns. Whatever your own views, and even if you reject this idea, failure to appreciate that others do not will limit understanding and debate.

The question in ethics of how we treat others can be usefully split into questions about how we treat others known to us and within our circle of everyday concern, and how we treat more distant strangers; the second person, and the third person. Briefly, although these may be collapsed into the dichotomy between self and other, they tend to arise in different ways and tend to need different approaches.

What this means for AI: In AI, the former set of questions may concern how we are changing how we relate to friends, family and colleagues through AI-mediated technology, and how we interact with robots; the latter, by questions such as the societal impact of technology, employment, taxation, discrimination in the use of algorithms. Indeed, the prospects of human extinction at the hands of AI raises questions of a different complexion again.

And the question of what if anything we owe the non-human world is raised in AI. For we are changing the world, AI will hasten these changes, and hence, we'd better have an idea of what changes count as good and what count as bad. Again, failure to appreciate the range of views on this question will limit debates.

2.8 What Counts as Adequate Justification and Argument in Ethics?

Consider the contestable nature of moral justification: We start from premises of uncertainty. We know that there is disagreement on questions of ethical value. Should we pursue happiness as the sole value? Should we care for all others equally? And so on. And we also know—by a quick perusal of philosophers' debates—that there is disagreement on the nature of justification in ethics, and what would count as a good ethical argument.

But the weight of ethical concerns means we can't simply put these questions in the 'too hard' basket. And it seems to be part of the nature of moral issues to require justification. The question 'why' always seems appropriate, especially if it comes from those affected by decisions. So, where ethical questions are concerned, we just have to solider on, somehow.

Moral Foundations Theory
Here are some related problems:

How to construct a code of ethics for AI, given that at least some of this AI will have global reach;

How to construct a code of ethics for AI that will be largely acceptable internationally;

How to embed ethical decision making and agency into AI: what ethics do we chose to embed in the machine, given variation in ethics cross culturally, and indeed, within societies?

(continued)

One way into examining these questions is to start from research into how people across the world actually do think about values. This in itself won't be enough, since ethics is normative, not simply descriptive, as explained earlier: it's no good pleading, 'but in some areas of London, gang culture holds that raping a rival gang leader's sister is a viable form of revenge, and they're increasingly finding that acid attacks are a cheap and handy response to insult', and leaving things at that.

Moral Foundations Theory is research that aims to understand what lies behind the variations in morality around the world (http://moralfoundations.org/). Researchers have probed moral views and claim that behind variation lies concerns that can be grouped in five or six main headings: care/harm, fairness/cheating, loyalty/betrayal, authority/subversion, sanctity/degradation, and perhaps additionally, liberty/oppression.

There will be societal and also individual personality variations in the emphasis given to these values. So note, that this has not found that 'really' humans have the 'same' morality at base. But it does indicate that there are common values, even if the emphasis is placed differently on these by some communities, belief systems, and individuals.

One important take-home lesson from this: understanding different ways of approaching ethical questions is the first step to seeing opposing points of view, and is a promising way to open dialogue with others.

2.8.1 How Do We Gain Moral Knowledge?

For some philosophers, this consists in gaining an appreciation of an independent moral reality. For others, it involves setting out one's moral goals (for example, the goal of maximising happiness and minimising pain) and then gaining the empirical knowledge to work out how best to do this in any given situation. For others still, morality is based not on objective reasons, but on subjective emotions. Such an approach will still be interested in conducting empirical inquiries, but these will be asking quite different questions. Others consider that we can best work out to do by considering the response of an 'ideal observer'. But is this someone stripped of all bias, of all emotion? Or someone who can see all biases, understand all emotions, and take them into account?

In drawing up codes of ethics for AI we need to assume certain broadly accepted notions concerning ethics. We can't just start from scratch. But it may be that in the very throes of discussing and implementing AI that some of the deepest disagreements about fundamental ethical issues bubble to the surface. We also need to consider how we can come up with the best, the most robust, the most workable set of guides and principles, given various disagreements about ethics, that pragmatically will attain assent and actually have a positive impact on action and outcome. And we need to think about how the process of arguing and debating all this needs to proceed.

2.8.2   The Elimination of 'Bias'

One of the first things people think about in improving moral arguments is the question of bias. It's now quite rightly routine, for instance, that conflicts of interest must be declared by participants in debate.

Eliminating bias in arguments seems an obvious goal, and some indeed hold out the hope that AI might help us to eliminate bias in ethical decision making. But what is 'irrelevant' bias? It can't simply be the presence of emotion, since (even if the views of those moral philosophers who place the basis of ethics in our felt responses to situations are rejected), in ethical judgement, it's often emotionally charged responses like empathy that help us to see what the moral issues are, and notice who's affected. Neither can it be any *simple* account of partisanship to one group, since a certain bias towards those who are suffering the most may be morally justified.

What this means for AI: This question is vitally important for the issue of who is involved in developing and implementing codes of ethics, as well as for projects to embed ethical decisions into machines. 'Getting rid of bias' may be a great goal, but to understand what it means, and how to do it, is another matter.

Algorithmic Bias in AI

There is increasing awareness that algorithms used to facilitate various operations can reproduce or create bias. This may be because the training data sets for the algorithms are themselves biased in some way, or because the operation of the algorithm itself creates bias. This will be an especially difficult issue where the AI involved lacks transparency.

But what is bias, anyway? Recruiters are going to favour the competent, other things being equal. Is this bias, if certain groups in society are less represented in the group picked out as most competent? There are a whole host of legal, political, sociological and moral arguments to be had here.

Okay, so we can at least start with the bias that law requires us to eliminate. But that's hard too. Here's just one of many potential problems.

A ruling by the European Court of Justice in 2011 has required that in order to eliminate the bias of gender discrimination in setting insurance policy rates, insurers must not give lower premiums to female drivers, (nor give men better pensions in view of their shorter life spans) (Kuschke 2012). But statistics show that women are in fact on average less likely to have motor vehicle accidents. Insurance works on precisely assessing risk. So any machine learning algorithm trying to work out premiums is going to end up finding proxies for gender. But, discriminating against a group indirectly through the use of proxies for a protected characteristic is also against the law. Ways to try to circumvent this problem include more and more personalised insurance calculations, such as reductions in premiums for drivers who install devices in their vehicles to track how well they are driving.

(continued)

But insurance is pooled risk. The end trajectory of highly personalised insurance premiums could be the end of insurance as we know it. Some people will have extremely small premiums, and some could well be priced off the road. Statistically, these will be disproportionately males, members of the very class who'd originally benefited from legal protection from discrimination with respect to motor insurance. However, on the bright side, pricing accident prone drivers off the road might be a relief to other road users.

Interestingly, one could point out that it's the very efficiency of the algorithms which has alerted us to the inherent difficulties created by changes in the law.

2.8.3 When Is Ethical Justification 'Finished'?

Ethical questions are often so complex that it's hard to make our answers exactly precise. But is there always a 'right answer'? Or are there some genuine moral dilemmas, where, whatever we do, there is some moral cost? It may be that we are sometimes faced with situations where different moral values clash, where they're *incommensurable*.

Why Is this relevant for AI? Where rapid technological and societal change is occurring which affects our relationships with each other and with the world, many of our values will be in flux. This makes it all the more likely that we won't have a fully worked out, coherent and consistent set of values. It's better to recognise this than to chase a false consistency. Witness current debates about privacy, an issue of particular concern in AI, where attitudes have developed significantly in relation to the use of technology, vary greatly depending on the context, and are also arguably internally inconsistent for many individuals (Nissenbaum 2004, 2010). An individual may value privacy in one area, while posting indiscrete personal information all over social media, and may see some data collection as routine, other data collection as a violation, but may lack consistent reasons for these distinctions.

Midnight Anguish and Slow Torment in Moral Reasoning
Especially where it's particularly hard to know what to do, and all the options have some pluses and some minuses, it's often noticeable that the subjective, felt quality of the decision making process is sometimes flagged as sort of place-holder for moral justification. 'Finding a decision particularly difficult to make' is sometimes accepted as a proxy for making a good decision. Watch out for this. It may or may not be something to worry about.

An example can be found in a report of an interview with Elon Musk and Sam Altman regarding the launch of OpenAI. This comes from a magazine write-up, so it's doubtless an incomplete account of Musk and Altman's own

(continued)

views of the matter; the example is meant simply to demonstrate the seductive idea that effort and difficulty indicates moral sincerity.

Interviewed on announcing the launch of OpenAI in December 2015:

Stephen Levy: I want to return to the idea that by sharing AI, we might not suffer the worst of its negative consequences. Isn't there a risk that by making it more available, you'll be increasing the potential dangers?

Altman: *I wish I could count the hours that I have spent with Elon debating this topic and with others as well and I am still not a hundred percent certain.* You can never be a hundred percent certain, right? But play out the different scenarios. Security through secrecy on technology has just not worked very often. If only one person gets to have it, how do you decide if that should be Google or the U.S. government or the Chinese government or ISIS or who? There are lots of bad humans in the world and yet humanity has continued to thrive. However, what would happen if one of those humans were a billion times more powerful than another human?

Musk: I think the best defense against the misuse of AI is to empower as many people as possible to have AI. If everyone has AI powers, then there's not any one person or a small set of individuals who can have AI superpower.

[(Levy 2015) (Emphases added.)]

The interview is interesting in many ways. There is an admission of uncertainty about whether OpenAI might increase the dangers of AI. But note how the opening proviso by Altman about the difficulty of the decision seems intended to provide assurance. Note, too, that this prolonged debate was said to take place between just two main people and an unspecified number of unknown others.

And note, too: There is however, a serious question to consider about what we are looking for in our moral decision making. In the context of AI, which focuses on speed, and which may operate using black boxes which no one fully understands, the reference by none other than Sam Altman to the slowness and difficulty of an ethical decision as markers of its probity, is telling. How machines operate, and how humans demonstrate the sincerity and integrity of their moral decision-making are poles apart on this account. Work on the psychology of time and decision making shows how different perspectives on the present and the future can affect conclusions and sometime distort judgements (Zimbardo and Boyd 2009).

2.8.4   Can We Necessarily Even Fully Articulate All Our Key Values?

Given the complexity and the importance of ethical questions, and given the social and technological changes being brought in by AI, it's highly likely that there are some profound values at play that we may find hard to articulate. We need to balance the demand to make our moral reasoning as robust as possible, with safeguarding against making it too rigid and throwing the moral baby out with the bathwater by rejecting anything we can't immediately explain. This point is highly relevant both to drawing up codes of ethics, and to the attempts to implement ethical reasoning in machines.

There is a good reason why we might not be able to articulate fully our most deeply held ethical responses. These may be more like the procedural memory we have for the deeply learned, automatic things we do each day that are driven into the fabric of our lives. There is a tendency among some philosophers to insist that the considered, articulated, coherent responses are the best, or the only ones allowable. But we would not dismiss as a fraud a concert pianist who could not explain precisely how their feats of virtuosity were achieved, finger movement by minute finger movement. Something similar might be occurring in our everyday and rapid moral reasoning.

And note it's our most fundamental values that are often hardest to articulate, for precisely the reason that these are the values from which we *start* articulation. The US Declaration of Independence (July 4th, 1776) states 'We hold these truths to be self-evident, that all men are created equal, that they are endowed by their Creator with certain unalienable Rights, that among these are Life, Liberty and the pursuit of Happiness.' Note the necessity of stating the self-evidence of these claims; this is a declaration of *faith*. No deeper ground of justification can be given. *"If I have exhausted the justifications, I have reached bedrock and my spade is turned. Then I am inclined to say: This is simply what I do."* (Wittgenstein 1973).

What's this got to do with AI? When we are trying to ensure that machines keep to our values, when we are trying to articulate those values in times of profound technological and societal change, we both need to be able to spell them out as rigorously as possible; but at the same time be aware, that the inability to do so may not mean that there is nothing of value there to be grasped.

2.8.5   Can There Be Such a Thing as Moral Progress?

There are various answers to this, from the optimistic answers of the utilitarian and social reformer John Stuart Mill (1863), to more pessimistic answers from those who see history as moving in cycles or just randomly. One lurking danger is a view that change is *ipso facto* change for the better.

What's this got to do with AI? In the context of AI and of technological change, one view is to see technological change as inevitable, and something we must adjust to but cannot realistically halt. It's useful to consider one's own assumptions about moral progress and social change. Excitement about AI often includes calls for its use in human enhancement. But in order to understand that something counts as enhancement in this context, we need to have a clear idea about what the desired end result is—and as should be apparent by now, that's still on *homo sapiens*' 'to do' list. Assessing and advancing moral progress, whether in individuals or in humans as a group, is highly complex (Wiseman 2016).

Transparency in Ethics and in AI—'What Plato Did'
Transparency in ethics has at least three aspects.

One is *visibility to others*. If others can see what you are doing, it makes it more likely you'll behave well. Philosophers have long known this. In Plato's *Republic*, Protagoras considered the Ring of Gyges, which magically renders its wearer invisible. Possessed of this, Protagoras argued, one would of course commit all manner of wrong-doing (Plato 1974). Conversely, much recent research lends support to the view that even *imagined* scrutiny by others helps us do the right thing (Zimbardo 2008).

The second is *comprehensibility to others*. Ethics demands a shared system of justification. In the *Republic*, Plato infamously argued that those in the top rung of society, the Philosopher Kings, dubbed the 'gold', had a grasp of moral truths but that the lower orders, or those dubbed the 'silver' and 'bronze' in society, were incapable of full access such knowledge.

And a related aspect is *accountability to others*. A corollary of Plato's views on knowledge and government is that, in governing those under them, the 'noble lie' could be justified to keep the *hoi polloi* in order. I take it that a view is abhorrent in any democratic society. It goes without saying that you can't claim to be adequately addressing ethical questions, if you refuse to explain yourself to rightly interested parties. Of course there will often then be a further question about who such parties are and what claims they have on you.

What this means in AI:

Firstly, The very complexity of much of AI means that there is often a particular question of transparency. If even its creators don't know precisely how an algorithm produced by machine learning is operating, how do we know if it's operating ethically or not? The frequently posed fears that without our knowledge we might be manipulated by powerful machines or very powerful corporations armed to the teeth with the opaque machinations of AI, gives a modern take on the Ring of Gyges myth. Only, now it's not actually a myth.

Having specialist knowledge, as professionals in AI have, does not entitle you to 'lie' to the people, nor to be in sole charge of questions that concern them; quite the reverse. Such specialist knowledge should mandate a duty to explain.

However, the question of how much transparency is legitimate in respect to certain activities is an open question. Only a fool wants the security services of their country to be fully transparent given the existence of real enemies; nonetheless drawing the line may be hard. Commercial companies also have reasons for secrecy. Which brings us on to the next point:

Secondly, there are many powerful actors involved in AI whose activities may affect billions of others; perhaps then, in some ways, a technological

(continued)

elite with access to arcane knowledge—AI professionals—are the new 'Philosopher Kings'. How they handle ethics, how they explain themselves, and whether they manage any system of accountability and dialogue, will be critical to any claim they might make to be truly concerned with ethics.

Some Notes on Disgust

We want our ethical arguments to be rigorous and we want them to be complete. But these aims may be in tension.

Some argue that some responses to moral issues are simply emotional reactions based upon what has been called the 'yuk' factor: an automatic response of disgust to an issue (Edmonds and Warburton 2010). This may seem like the reaction of someone uneducated in restraining their thoughts and submitting them to the test of reason. Those who warn against basing views on 'disgust' may find support in experimental evidence indicating that manipulating disgust responses can alter moral judgements (Haidt 2013).

But, psychologists have argued that disgust reactions track a kind of immune response to protecting the self, the community and its boundaries. Disgust reactions are linked to notions of sanctity or purity. Work on the range of values in moral psychology shows that those with certain political views (broadly, liberals) tend to focus on a narrower range of moral values than those with opposing views (broadly, conservatives). The latter include values of sanctity and purity which may result in responses of disgust (Schnall et al. 2008).

But, it is among those philosophers who themselves tend to argue for a narrower range of values (autonomy, welfare, justice, for example) that the arguments for eliminating considerations of disgust (and dignity) can generally be found. So, are these philosophers simply more rigorous in their quest for moral justification? Or are they more limited in their appreciation of a range of values?

What's this got to do with AI?

One: Some of the ethical questions in AI concern how we should delineate the boundaries between humans and machines. So, we should expect that some responses to some possibilities will involve disgust (for example, calls for the development of post-human cyborgs).

Two: Since we know that different groups of people see such reactions as relevant to ethics, or as irrelevant to ethics, this has implications for how we constitute our discussions of AI ethics.

Three: Interestingly, as mentioned, disgust responses are linked to notions of sanctity or purity. Those calling for the removal of consideration of disgust, or 'woolly' notions like human dignity, from discussions of ethics, are themselves, of course, exhibiting a variant of a call to purity.

2.9   Moral Relativism, Moral Justification and AI

How can justification of ethical arguments proceed, given that there is a large variety of moral systems and ethical beliefs, not just within a society and culture, but between different cultures?

Why is this an issue for AI? Because many forms of AI, by their very nature, affect people across societal boundaries. Because AI is predominantly being developed in certain parts of the world. Because AI, along with other technologies, is helping to connect individuals and groups from different social and cultural groups.

What should we do about it? Again, a book of this length cannot hope to answer the question. But we need to be aware of the questions. Communication, open dialogue and debate, and diversity in participation, go some way towards recognising the issues.

Note too that there are many responses one might take to moral diversity around the world. Recognising differences between cultures in moral codes, and valuing the contributions from a variety of cultures, has not led all to conclude that moral beliefs are simply relative to different societies.

Having Your Relativist Cake and Eating It: *Not Such a Good Idea*

Here is a commonly expressed argument behind a particular view of moral relativism:

PREMISES: Morality is simply the expression of socially constructed value judgements. Other societies have their standards of judgement, we have ours.

CONCLUSION: Therefore, we should not judge other cultures.

Such a view is often motivated by the finest principles—concern not to condemn what we don't understand, and concern for power imbalances between the wealthier and the less wealthy. There are many examples where havoc was wrought by 'interfering' in other cultures. And we have much to learn from dialogue with others.

However, this view involves taking what the philosopher Bernard Williams once described as the 'mid-air' position (Williams 1976). The premises state that all value judgements only make sense *relative to a social system*. But the conclusion—a value judgement—is announced as if it is some *universal truth*.

But, if morality is *always and only relative to societies*, from what society do we judge that it's wrong to judge other societies? From some 'mid-air' position, outside of any culture, from which it is possible to pronounce universal truths? But … I thought you said all value judgements only make sense from within some society or other?

Moreover, such a simplistically sketched view may rest on an assumption of a series of isolated and homogenous societies which each contain their own autonomously created set of values. This is a greatly simplified view of the

---

(continued)

complex world we face today, and raises specifically difficult issues in regard to international issues. The possibilities that AI itself brings are indeed helping to further create and disrupt links between cultures and to disseminate information.

Furthermore, such crude cultural relativism tends to present individual societies as harmonious clubs where everyone agrees on the presiding values. But this is not true of many societies, and perhaps true of none. There are almost always some groups in society whose views are not adequately heard, and whose interests get short shrift. Moreover, taking notice of such people is of the very essence of ethics. So, this commonly held form of relativism may end up doing the reverse of what the often well-meaning people behind it wished to do—it may end up supporting the dominant views of the most powerful people in other cultures.

And note this complexity. The currently dominant views of morality in Western thought are universalist in nature. This is behind moves like the Universal Declaration of Human Rights. But, if 'our' morality is universalist in nature, then, from a relativist view of morality, who can argue that we should not be universalist? So, paradoxically, if we maintain a crude moral relativism, then there is reduced ground to argue against imperialist expansion or a global takeover of systems of AI.

What does this mean for AI? AI crosses national and cultural boundaries. We need to think about how we develop a robust ethic which addresses this without simply degenerating into a 'pick and mix' approach, where if someone else wants to use AI to instigate, say, the total surveillance of their population in an attempt to fine-tune brainwashing, we *simply* say, 'oh well, each to their own'. This is a crude example; the point is how hard it is to draw a line between praiseworthy respect for other cultures, and turning a blind eye to moral wrongs.

2.10 A Distributed Morality?

Note that calls to end bias, and many notions of justification in ethics, often rest upon an assumption that there is one thing that it's right to do, and that this is the same for all agents. But many argue, often on the basis of research in various branches of the social sciences as well as in philosophy, that morality is (at least sometimes) socially distributed, so that differently placed actors within a situation have different moral roles to play; and that *this is better than a 'homogenous' morality*. As ever, there are variations of detail in how a distributed morality might be understood (Floridi 2013; Floridi and Sanders 2004).

What does this mean for AI? It has of course implications for the responsibilities of individuals and teams in AI, for questions about autonomous systems, including systems involving both humans and machines, and for questions around building in ethics into intelligent machines.

2.11   Moral Agents

What is it to be a moral agent, what motivates us to act morally—and what prevents us from acting morally? On some ethical theories, all that matters is that the best result obtains. Such accounts are neutral with respect to agency; it doesn't matter who acts, so long as the job gets done. On others, agency matters, and matters crucially in a multidimensional way. Deontological and virtue ethics theories take such a line. This is a fascinating and complex area that has received intense debate and scrutiny. Here are a couple of pointers for why this matters for AI, and for developing codes of ethics in AI.

If agency does not matter, then we can outsource our moral decisions and actions to another competent person, or even to a machine. But on the most plausible views of ethics, the intention with which something is accomplished makes a difference to its moral assessment, and it matters who it is who is acting, and why they act as they do. Even consequentialists usually see the point of the questions they are asked about the place of agents in their account of ethics (Scheffler 1988).

So, what does this mean for AI? If our actions are mediated by a machine which lacks transparency in some respect, how do we ensure that they are ethical? Suppose I used an algorithm designed by machine learning to make a policy decision. How can I be held accountable for decisions made in such a way? On some views of ethics, well, never mind, so long as the outcome is okay. On others— not so fast.

But note that addressing such highly complex questions can mean examining the basis of claims of agency and autonomy, in ourselves as well as in machines. And in part, much work on the development of intelligence and agency in machines is examining the nature of intelligence and agency in humans. This means that we might perhaps upset the philosophical applecart on which certain views of ethics rest. For instance, are we using ideas of moral agency which assume humans have free will?

This debate is far too interesting to pursue in great detail in this little book. But, note how again, how deeply questions about AI go when we think of them alongside questions of ethics. In drawing up codes of ethics for AI, it will be important to examine what assumptions are being made about moral agency.

2.12   Moral Motivation

There is also the question of *moral motivation*. Authority, and motivation to adhere to codes, may stem from 'soft' powers such as the respect for the originating body, or for the colleagues or the process by which codes of ethics were drawn up and discussed. For those assuming that so long as the very clever people who work in AI produce codes of ethics, this will be enough to inspire confidence in those codes, some humility may be found in personality research which indicates that there is no correlation between how intelligent you are, and how likely you are to follow codes of conduct. Psychological studies have found a null or negative correlation between IQ and the trait of conscientiousness, which roughly translates as 'character' (Luciano et al. 2006; Moutafi et al. 2004).

What's this got to do with AI? Let's face it, AI is run by people who are generally pretty bright, at least in certain ways. But it's a mistake confidently to presume that clever people will draw up, and implement, good codes of ethics, simply in virtue of their intelligence.

2.13 AI, Codes of Ethics and the Law

There is a strong and complex relationship between ethics and law. Codes of ethics are nested within the appropriate legal jurisdictions of local, national and international laws, and seek to adhere to these. However, especially when technology is rapidly advancing, the law might not be able to keep up, and professional bodies and others considering ethical aspects of that technology might well lobby for appropriate changes to the law. It may be possible to amend codes of ethics issued by professional bodies more flexibly and more rapidly than national, and especially international, laws.

There may be great differences in some aspects of the law between different jurisdictions, some of these being differences of great relevance to AI. For example, there are significant differences between the laws on data protection and privacy in the US and in Europe, which can potentially be highly relevant to codes of ethics for regulating AI, and indeed, to how AI is developed.

Meanwhile, how can technology cope when a legal regime might be a stumbling block to its development? For example, legal regimes may be rightly concerned about the development of autonomous vehicles, yet this might slow the development of technology which in the longer term could have a beneficial impact on road safety.

One possibility is to test technology in more permissive jurisdictions. One problem might be certain countries paying a price for the development of technologies from which other countries are more likely to benefit. Suspicion has been raised that testing for paediatric medicines may take place in less developed or developing countries where children are not so vigorously protected (Gulhati 2005). Another more attractive possibility is to have prescribed certain areas where experimentation with technology was permitted, subject to improved regulations (Pagallo 2011).

Law has to be applied, and applied rigorously and consistently across a wide range of circumstances. Attention to how the law might be updated to accommodate various developments in technology, including AI, may proceed with an attention to detail from which ethics could sometimes benefit. Contrariwise, close attention to legal judgement in relation to AI as it unfolds in case law can be both useful for considering ethical issues, and important to note for critical commentary as thinking in AI unfolds.

For example, the 2016 decision in State v Eric Loomis (State of Wisonsin v Eric Loomis 2016) concerned whether the use of the COMPAS algorithm in determining sentencing was fair or whether it violated the Constitutional right to due process. The finding was that it was used appropriately. A legal decision such as this will make reference to precedent and law in the appropriate jurisdictions of course. There can naturally also be broader debates about whether such legal decisions really do capture 'fairness' in such cases. Indeed, in this case, Loomis filed a petition for the writ of *certiorati* concerning the judgement; in an unusual move, the Supreme Court of the US ordered the State of Wisconsin to respond, and on March 6th 2017 in an even more unusual move, the Supreme Court issued a CVSG, a call for the views of the Acting Solicitor General (Admin 2017). This reflects the gravity of the concerns about the lack of transparency in the use of such algorithms and the possible threat to procedural justice and fairness. This level of scrutiny by the courts is to be welcomed and is indeed necessary with the introduction of AI which is potentially altering fundamental tenets of our legal system.

Additionally, the very fact that there are sometimes important relevant differences between jurisdictions on the law, which then shapes debates about ethics and codes of ethics, means that examining the possibilities of different legal regimes can be a good way of thinking more laterally about what is possible and what kinds of legal reform might be desirable.

Chapter 3
Does AI Raise Any Distinctive Ethical Questions?

Abstract In developing codes of ethics for AI, it's important to consider how the ethical issues concerning AI relate to other ethical issues. Some of the questions overlap with those relevant to other rapidly developing technologies, especially where those technologies involve questions about how we humans understand ourselves and our place in the world. Issues that concern both AI and other areas of science and technology include the difficulty of considering ethical issues where predictions about the future may be difficult to make; where technology is embedded in complex ways in society, the economy and culture; where the analysis and use of large amounts of data are at issue; and where complex, large systems give rise to difficult questions about the distribution and attribution of responsibility. Characteristic ethical questions regarding AI concern its typical enhancement or replacement of human agency; crucially, questions of agency are at the heart of how we understand ethics. AI, like other technologies, is often hyped up; it's important to see through the hype to as realistic a picture as possible. Hype can distort our ethical thinking in dangerous ways, and some of the key hazards of hype regarding ethics are outlined.

3.1 Methodology: Focusing in on Ethical Questions

The ethical questions of AI are sometimes presented as if they are new and unique. But it's clear that some of the questions that AI raises are common to other concerns, such as other developing technologies and changes to economic practices. Understanding the links with other issues is important methodologically, for this can alert us to common ways of thinking that can be drawn on in moving the debate forward. But it's also important to consider if there are any distinctive ethical questions in AI, since if we are faced simply with a list of very broad ethical issues which are not focused upon the specifics of AI, we may end up with codes and principles which lack sufficient grip on concrete circumstances to achieve substantive impact. There will always be arguments about whether certain features are unique to AI. So, here, I consider if there are any features which are characteristic of AI, and which have significant ethical import.

© Springer International Publishing AG 2017
P. Boddington, *Towards a Code of Ethics for Artificial Intelligence*,
Artificial Intelligence: Foundations, Theory, and Algorithms,
DOI 10.1007/978-3-319-60648-4_3

3.1.1   How Do We Identify Ethical Problems as New?

Could certain applications of AI render more visible to the clear light of day, ethical issues which had been there all along? Perhaps this might be by magnifying the speed and efficiency with which certain tasks are done. Perhaps this might be by the reach and power of its wider effects e.g. on social organisation and on economics. Perhaps it might be because AI may articulate step by step what is needed to simulate human reasoning, and hence in building AI, we might become more acutely aware of the capacities and limits of our own reasoning, and perhaps more acutely aware of how we think about issues with moral implications. For all such reasons, and possibly more, AI might discover, as well as perhaps create, ethical issues for us to ponder.

Regulating Medical Robots: Is this an Issue for AI, or an Issue for Medicine?

The European Parliament's report on European Civil Law on Robotics (European Civil Law Rules in Robotics 2016) discusses the question of how precisely robots are defined. Legislation needs clear definitions. And robots come in a variety of forms. Robots may not involve AI as such.

One issue is whether we treat robots which don't have autonomy differently from those that do. The report points out that medical robots fall into the 'master/slave' category of robots, since they remain under the control of the surgeon or surgical team. 'Nevertheless, the European Union absolutely must consider surgical robots, particularly as regards robot safety and surgeon training in robot use.' (p. 9)

But note that practice of medicine, surgeon training, and medical devices are already subject to close scrutiny and regulation. In addition to law and professional regulation, there are also many active patient interest groups who concern themselves with such matters. Medical robots rightly deserve ethical and regulatory attention, but there is no particular reason—certainly not one to be found in this report—to suppose that existing systems for scrutiny and response in medicine are not adequate to the job. The absence of autonomy in these 'master/slave' robots is relevant to reaching such a conclusion.

3.2   Many Ethical Issues in AI Are Shared with Other Rapidly Developing Technology

Here are just a few.

Problems of prediction: The very fact that AI is a rapidly developing technology means that it's hard to predict what will occur even in the near future. As well, it is hard to anticipate ethical questions in advance and to produce codes and regulation as fast as the AI develops. But this is the case in other areas of science and technology as well.

Interface with social and cultural issues is an issue in all technology and AI is no exception. In considering ethical issues we need to consider how there may be subtle, yet pervasive, social, cultural, and economic impacts from the use of AI.

The manipulation of data: Some issues, especially in certain areas of AI, concern the use and manipulation of masses of data, and this issue arises in many other areas too. There has been work for some time now on ethics of data management, for example, in genomics and in other medical research.

Complex systems and responsibility: AI often involves highly complex systems, with those working on different elements unknown to each other, and the effects of the operation of AI may be some distance from producers, for instance when machine learning produces algorithms which may affect millions of others far distant to the source. Questions about how responsibility works in such large networks of researchers is already receiving attention (Boddington 2011).

Nonetheless, there are some characteristic ethical challenges in AI.

3.3 Ethical Questions Arise from AI's Typical Use to Enhance, Supplement, or Replace the Work of Humans

It is difficult to give a precise characterisation of how the extension of human agency by AI differs from that offered by simpler automation. What is characteristic of AI is not just that it extends or enhances human agency; nor that it extends or enhances human reasoning, for something as simple as an abacus does this. AI characteristically enhances or replaces human decision-making and human judgement. It may enhance or replace human action and/or human perception, and, may attempt to simulate human emotions.

The extension, enhancement and replacement of human agency and reasoning in AI serve as the loci of many of the ethical issues that arise in its use, sometimes presenting us with vivid versions of old questions. Note that these will vary depending on the precise application and context, and on how far-reaching are the effects of the AI. In some instances, AI does what humans could do, if they had enough time; in other instances, it seems to enhance human agency and reasoning to provide calculations and perform actions that we simply could not do unaided. Some recurring issues arise.

AI and transparency: Some applications of AI give us answers that are in principle hard to check, and where it's hard to know how the answer was even reached, as when answers are produced by machine learning. There's currently debate about such questions, about whether, for example, we'll ever be able to train machines to explain their otherwise opaque decisions to us.

AI and the control problem: The more powerful AI is, as a general rule, the less we are going to be able to control it. There's no easy way to characterise this, and it's always going to be necessary to look in detail at specific cases. In some instances, autonomy might give more control—an autonomous missile can be programmed with a more precise goal than a bullet which once released, is out of our control. But what tends to be characteristic of AI is autonomy of judgement, of decision, of action; and as we have seen, these are key concepts driving accounts of morality at a very deep level.

This means also that work in AI itself is directly tackling issues which are also of central concern to ethics. Artificial Intelligence is perhaps unique among engineering subjects in that it has raised very basic questions about the nature of computing, perception, reasoning, learning, language, action, interaction, consciousness, humankind, life etc. etc.—and at the same time it has contributed substantially to answering these questions (in fact, it is sometimes seen as a form of empirical research) (Müller 2012).

AI then tends to give rise to questions that we see of particular significance to ethics, and to our values more broadly. As we've seen, at the heart of any account of ethics is an account of moral agents, their powers and their limitations. And AI is precisely about the extension of our powers as agents. One issue of particular interest is how in thinking about this, we have a tendency to idealise both the agency of humans, and the agency of machines, and this will be discussed further in Chap. 7.

Robot Camel Jockeys: 'Pimp My Ethics'

Robots were often employed to replace child camel jockeys in various Gulf states where camel racing is a popular sport. The use of child camel jockeys was highly problematic: there is considerable evidence that very young children were taken from countries such as Bangladesh, Pakistan and Sudan, deprived of food to keep their weight down, treated very harshly, and subject to the frightening and dangerous task of riding camels, all for the sake of a sport. This practice was also against laws on child labour, yet continued despite these laws. There are widespread reports of sexual abuse as well (Gluckman 1992; Lillie 2013; Pejman 2005).

Firms created small and simple robots to replace the child jockeys. This has been hailed as a success; they have even been described as 'the robots that fight for human rights' (Brook 2015) and that 'there are some issues that really can be solved with innovation and technology' (Rasnai 2013). Although these robots are so simple that they scarcely count as AI, it seems then that this is an instance of robotics put to great moral benefit. There have been positive write-ups in various tech magazines, and Vogue has even produced a glamorous photoshoot of camel racing with robot jockeys (Shaheen 2017).

But this case also serves as an example of how attention to even this simple technology can divert attention from other issues.

Hailed as an ethical 'win-win', reading further down some of the tech magazine articles on this topic shows that the situation is more complex, and

(continued)

further research demonstrates this too. There are difficulties in repatriation of the boys (Pejman 2005), and one organisation estimates that 3000 of these boys have simply vanished (Ansar Burney Trust 2013). It's not even entirely clear if the practice of using young boys in appalling conditions has been completely eradicated (Peachey 2010). So, has the rush to praise the use of tech to solve a moral issue been premature?

There are moral pluses and minuses of the allure of technology. Note that one reporter describes how the robots were adopted because they were 'cool': 'the message is clear: pimp my camel' (Schmundt 2005). If we take this analysis at face value, it means that the very coolness and allure of technology served a good moral purpose, of helping to stop the use of boy jockeys where laws requiring this had not. That seems good, doesn't it?

But note then too, that if the camel racers and owners merely used the robots instead of the boys because they thought they were 'cool', this does not suggest that they were motivated by a moral awareness of the devastating impact on the lives of the child jockeys.

3.4 We Also Need to Consider the Methods of Production of AI

The development of technologies in AI gives increasing possibility for a few individuals, corporations or research groups to make significant advances with scant oversight, notwithstanding the current efforts of many to consider the ethical and policy issues involved. Work can be done by those working entirely outside the framework of any professional accreditation.

And there is an increasingly steep gradient of wealth and resources for those accumulating capital in IT in general including AI (Piketty et al. 2014). Again, this is not unique to AI, but is characteristic of much IT in general, and the wealth, and technological reach, of some of the big players in AI is so large that realistically, this is a major feature of the world economic and, indeed, political landscape. All these issues present challenges in considering the development of codes of ethics for AI, especially when coupled with the breadth and power of AI to affect the lives of people globally.

3.5 Hype in AI and Implications for Methodology in Ethics

It is easy to be dazzled by the myriad claims made about AI and the scent of hyperbole that often surrounds it. Hype can concern both the technology itself and the ethical issues it may raise. The hyping of technology and its potential dangers and benefits tends to occur particularly when notions of speed, rapid technological change, and money, are present; for example, the presence of hype in genomics research is well established (Caulfield and Condit 2012). The framing of a technology as having a fundamental impact upon humanity and on our self-image, as has also happened with genomic technology, is also at play in many discussions of AI (Joy 2000); indeed, such claims may be true, making it all the more important that they are examined closely.

So, is the current attention to the ethical issues in AI a hyped 'moral panic'?

Work in the analysis of rhetoric and language demonstrates how the framing of an issue, and precisely how its ethical problems are described, impacts upon our responses (Fischer and Forrester 1993; Majone 1989; Throgmorton 1993). Problems will be made sense of and acted upon through descriptions which frame the discussion, highlighting certain aspects, and making certain contrasts (Rein and Schon 1991; Hajer 1993).

Consider the biotechnology industry which has also been the target of multiple moral attacks (Newton 2001); these are not necessarily grounded in realistic assessments, with hard-to-explain regional differences in publicly-voiced concerns (Ford et al. 2015). Hype may be used, sometimes knowingly, to fulfil a particular purpose. For instance, accounts of synthetic biology lurch between hyping up how new it is, typically in the context of attracting investors, and insisting that it is a well-established field only doing what has been done for millennia, thereby downplaying the need for specific regulation (Parens et al. 2009).

Likewise, we are warned about the moral dangers of AI, and also the moral dangers of clamping down on progress in this technology. Hype about future possibilities can distract us from present realities (Crawford and Calo 2016). For instance, hype about a possible malevolent superintelligence being developed some time in the future could distract from the need to consider ways in which AI is already affecting our everyday lives now. Any work in the ethics of AI then needs to think carefully through the hype and the reality; while taking into account that the very uncertainty of the field and of how it might develop makes it hard to work out what is hype and what is realistic.

3.5.1 Hype Can Both Distort Our Ethical Reasoning, and Reveal Things of Potential Interest

Questions in ethics require close attention to all aspects of a situation that are morally relevant, and must weigh these against each other, a difficult task requiring close attention. We need to be aware of how our attention is being commanded, and consider if we are being drawn to something of prime relevance, or if we are being blindsided. There is no easy answer to this, and it's one of the reasons why ethical debate and dialogue with others is vital (Bazerman and Tenbrunsel 2011).

For example, let us examine a claim in a widely-read article that states issues involving AI present 'the key question for humanity today' and that artificial intelligence holds unparalleled promise because 'everything that human civilisation has to offer is a product of human intelligence' (Hawking et al. 2014). The last remark is a loose and inflated view of the power of human intelligence. Note the ambiguity of 'product of human intelligence'. This could mean merely that human intelligence had something to do with all the good things of civilised life. But it seduces us to consider that 'everything' of value in civilisation is the designed result of the application of intelligence, and it may lead to hubris, if we consider that nothing is due to serendipity or other factors.

Moreover, human beings are not simply characterised by intelligence but by sociability. It's common in the context of AI to read casual remarks about how intelligence is what has brought humans so far, such as this comment from Guruduth Banavar: 'From the evolutionary point of view, humans have reached their current level of power and control over the world because of intelligence. . . AI is augmented intelligence' (Conn 2017a). Such remarks often slide into claims that it will be, therefore, artificial *intelligence* which leads us forward to the future (Kurzweil 2001). However, although some claim that it is human intelligence which has formed the impetus behind our evolution, other evolutionary scientists beg to differ: our sociability, pair-bonding, and religion, are also prime candidates to explain our extraordinary success (Barrett et al. 2002; Zauzmer 2017). And with regard to developments in AI, some sensible humility about the impacts of human 'intelligence' on civilisation, and a measured appraisal of our abilities to control AI, is precisely what's at issue. Yet the phrase seduces us into a world view that breathes the promise of human control over what we do. Such tunnelling of thinking towards intelligence alone can be characteristic of much talk about AI, and can have a real impact on how ethical issues are discussed.

3.5.2 Hype About AI Can Channel Our Thinking About Solutions

A focus on the dangers of AI, coupled with optimism about its potential, can lead to an overreliance on AI as a solution to our ethical concerns—to an 'AI on AI' approach. Problems are seen as solely technological, hence as requiring technological solutions. But creative thinking can often produce a wider range of solutions to problems. The common view that 'technical cause = technical solution' is not necessarily valid, just as it is not necessarily valid in other areas such as medicine or psychology. Bear such points in mind when the question of how we assess the harms and benefits of something as complex and as deeply embedded in our life as AI.

Critiquing the Core Values of a Profession

The practice of modern medicine has saved and extended many lives, and improved the health of many (along with other developments such as plumbing, electricity, widening education, and the mechanisation of agriculture), so it's easy to take its value for granted. A set of professional regulations is unlikely to critique the core activities of the profession as such. Such critique tends to come from those outside the profession, such as groups representing public concern, critical social scientists, or mavericks within the profession. Yet such critiques can be very helpful. Often, giving a name to concepts helps move debate forward.

Healthism is the notion that health is the supreme value that should be pursued above all others; this can be questioned (Fitzpatrick 2001; Crawford 1980).

The concept of medicalisation has been influential in debates about the benefits of medicine. *'Medicalisation'* is the process by which phenomena are analysed in medical terms, with solutions sought in medical terms, where this represents a narrowing of possible options (Illich 1976). For example, to treat 'low mood' as only a medical issue, and with a pharmaceutical response only, may in some cases overlook the root causes of an individual's problem and may mask helpful solutions.

Geneticisation refers to the same phenomenon, applied to genetic explanations of conditions (Lippman 1991), where this blinds us to other aspects of the causation and treatment of disease and health.

Likewise, *iatrogenesis* names the phenomenon whereby the practice of medicine actually causes problems, which are then treated with more medicine (Illich 1976).

Virtually everyone is against eugenics. But many practices in modern medicine may be described as *liberal eugenics*, where the state does not mandate the practice, but actions by individuals under the banner of individual reproductive choice occur which change what kind of children are born. Some hail this; others note the strong arm of the state may be replaced by other pressures implicit in much modern medicine. Either way, the label 'liberal eugenics' is useful for catalysing debate (Agar 2004).

Are there equivalents of medicalisation and iatrogenesis happening in relation to AI? *Artificialintelligenciation* is a bit of a mouthful which I freely admit may not make it into the Oxford English Dictionary. But perhaps we ought nonetheless to look out for it. And perhaps codes of ethics ought explicitly to consider this too.

And where AI concerns the 'enhancement' of humans in various ways, the questions raised about how our thinking may be tunnelled concerning the benefits of medicine may be of direct relevance.

3.5.3 Impacts of Hype on Moral Thinking

Here are some of the ways that hype can impact how we think:

Distortions in thinking: On occasion, discussion of AI can simplify, exaggerate and idealise what it is to be human, and what it is to be an intelligent machine. Hype may not simply and harmlessly draw attention to the ethical issues, but may distort the very questions that we need to address. Hype can shape the (dis)appearance of humanity in how we perceive technology, the relationship between humanity and machines, and the attenuation of our visions of humanity. We will discuss the idealisation of agency in thinking about AI and ethics in Chap. 7.

Hype, ethical dangers, and public image: One important potential impact of hype is that fear of being branded one of the 'bad guys' may lead to individual or collective attempts to promote oneself as on the side of the angels. This might be for intrinsic or explicitly strategic reasons. There are serious dangers for institutions which fall for this trap.

At its worst, appearing to be ethical might trump concerns about being ethical. Hyping the moral dangers of AI might produce what has come to be known as 'virtue signalling' (Bartholmew 2015), where an individual or organisation proudly proclaims their ethical credentials in ways which acts as a substitute for actual action (Hogben 2009). Content then might perhaps be simply empty displays of ethical probity ('we are passionate about the future of the human race', and so on).

Hype may have a tangible impact upon the content of codes. The rush to go address prominent ethical issues may lead to foregrounding one moral value, possibly at the expense of others; the ethics is set to 11, to paraphrase the film *Spinal Tap*; this has been dubbed 'ethical enthusiasm' (Hammersley 2009). For example, concern for data privacy might create conditions that make valuable research onerous or even impossible.

Emphasising one moral value without attention to balance has particular hazards. Aristotle long ago argued that at the extremes, virtues turn to vice (Aristotle 1999). Research in moral psychology and personality traits corroborates this; personality traits spread out along a normal distribution, which indicates that there are disadvantages of inhabiting the extreme ends of the distribution. For instance, the trait of empathy at its extreme can lead to overlooking hazards to self and other in concern for those who are the object of that empathy, and aggression towards any perceived threat to the object of empathy; yet aggression in some circumstances is very valuable. Values need to be balanced with each other, against the realities of a particular context. Hyping up the importance of one value, to the exclusion of others may occur when panicking about the impacts of AI.

Emphasising One Ethical Value: Excluding the Less Well Resourced

Policies designed to produce good ethical practice are sometimes easier to comply with, if you are well resourced. For example, the data sharing policies of many scientific research institutions include clauses that data sharing should be encouraged, by sharing data only with those who enter into reciprocal agreements—who are dubbed 'good data sharers'. But, those from less well-resourced institutions may have less to share, or may lack resources to guarantee certain standards of data curation. The least privileged researchers may end up with the worst deal, unless some special provision is given (Boddington 2012).

But then, focusing on one highlighted value may indeed be what an organisation, on reflection, decides to do.

Here's an example from AI. The Icelandic Institute for Intelligent Machines (IIIM) Ethics Policy for Peaceful R&D eschews military funding, and has rules about collaboration based on this:

2.3 IIIM will not collaborate with any institution, company, group, or organization whose existence or operation is explicitly, whether in part or in whole, sponsored by military funding as described in 2.2 or controlled by military authorities. For civilian institutions with a history of undertaking military-funded projects a 5–15 rule will be applied: if for the past 5 years 15% or more of their projects were sponsored by such funds, they will not be considered as IIIM collaborators (IIIM 2015).

Researchers who are not in the position of finding alternative sources of funds will therefore be excluded from potentially beneficial collaboration.

This is not a commentary on the rights or wrongs of the IIIM's stance; the IIIM is perfectly free to set their own policies. These issues might however be a central concern for government agencies, or for international professional bodies, who probably would not wish to disadvantage any members, least of all those are already struggling with resource access.

3.6   Conclusion

I have argued that thinking about AI will require us to think clearly and deeply about some fundamental questions in ethics. AI and ethics are intimately concerned with fundamental questions of agency. Various questions in ethics concern how we relate to friend and stranger, to the world at large, and how we treat ourselves. AI promises to make changes to all of these dimensions of our lives. And we need to consider how far AI has common concerns with other areas, and how far it raises distinctive questions, while being careful not to fall for the distorting effects of hype upon our moral thinking.

Our task then is this: we need to develop codes of ethics in a situation of uncertainty, derived not just from the rapid development of technology, and not just from the diverse views about this technology, but from the rapid technology which is changing potentially how we relate to others, to the world around us, and our sense of self—in other words, impacting upon the very foundational basis of any ethic.

But here, we are not looking in general at the ethics of AI. We are concerned with developing codes of ethics for AI. I turn now to consider the essential elements of professional codes of ethics, before discussing how AI challenges these elements.

Chapter 4
Codes of Professional Ethics

Abstract This chapter outlines the features of the professional practice which lead to the necessity for codes of professional ethics, and which underpin the nature and typical content of such codes. There are a variety of codes and regulations regarding professional practices, which may serve different purposes. Members of a profession possess certain skills, knowledge and capacities that their clients and the general public typically lack. This creates a gradient of power and of relative vulnerability between the professional and others. Codes of ethics aim to mitigate the potentially deleterious effects, or the misuse, of such professional power. Codes of professional ethics may be backed up by hard or soft power. Since each profession deals with a certain area of endeavour, codes of professional ethics typically concern themselves with values, benefit and harms in relation to their own area of expertise. Nonetheless, there are general values underlying such codes, even if these are implicit. These may be hard to articulate and may indeed be controversial. The value of autonomy is examined as an example especially relevant to AI. Codes of ethics can only function effectively with both adequate institutional and societal backing. Understanding the history and context of development of codes of ethics is important to understand their underlying values, and especially where social and technological change is occurring. Codes of ethics may develop in response to catastrophe, in anticipation of problems, and with reference to codes of ethics in key areas, and all of these may give rise to problems. Codes of ethics may have certain failings, and in some cases even make a situation worse.

4.1 Introduction: The Varieties of Ethical Codes

There are different possible formats for ethical codes, regulations or guidance. These include codes of professional conduct produced by various professional bodies for their members, or by other regulatory bodies; safety standards, often produced by industry or governmental bodies and possibly by statutory powers; and research ethics codes and regulation produced by institutions funding, carrying out, or overseeing research. There are also statements produced by prominent members of a profession, such as the Asilomar Recombinant DNA Principles (Berg et al.

© Springer International Publishing AG 2017

P. Boddington, *Towards a Code of Ethics for Artificial Intelligence*,
Artificial Intelligence: Foundations, Theory, and Algorithms,
DOI 10.1007/978-3-319-60648-4_4

1975), or by special interest groups, such as those opposed to autonomous weapons. Discussion papers and interim guidelines may be especially relevant in disputed areas or where technology is developing rapidly. Codes or guidance may be enforceable by law or by other penalty, such as by deregistration or professional disciplinary action, or may offer guidance only.

The project from which this book arises focuses on professional codes of ethics for artificial intelligence researchers. But given the turbulent landscape in which AI is developing, professional codes of ethics will need the backing and support of other formalised or less formal, and institutionalised ways of addressing the ethical questions confronting us. However, for simplicity and ease of explanation, we are going to commence by considering professional codes of ethics, their typical purpose and nature, and then draw out implications for codes of ethics for AI, as well as more widely for how professional and public debate in this area should proceed.

4.1.1 The Purposes of Codes and Statements of Principle

The various codes, declaration of principle, regulations, and laws that exist can have complementary roles, and may differ from each other perhaps because of matters of substance, and perhaps because of their context and intent. There is a role for codes and sets of principles that are aspirational. And there is also a complementary need for codes which can be operationalised into concrete action; this is especially the case where codes of ethics are intended for guidance for engineers at the front line of developing AI.

Codes and regulations in different settings may not translate well to other settings; or they may be very useful for cross-fertilisation of ideas. Codes may be designed for local or national use, or may aspire to international application. A commercial organisation will have its own financial interests which may be nested within legal and ethical concerns, but which will have an impact upon any codes of ethics they produce; government codes and regulations may deal extensively with economic issues but with quite a different agenda than that of a private corporation or a professional body. In a legal context, there are ethical considerations in formulating and applying the law, but the law may lack the nuance that is needed for a rich account of ethics. Contrariwise, the law needs to spell concepts out in sufficient detail that judgements can be made in particular cases. This can mean that the law, including case law, can be a very useful source for considering how to operationalise and add detail to general and abstract concepts in ethics. This could be particularly useful in our area of concern, where developments in technology and changes in social relationship are presenting us with the need to apply central ethical concepts in new contexts.

A code of ethics should not be seen as complete and self-sufficient, for such codes exist in a particular context (Bowden and Surma 2003), and without the backing of a supportive institution, a code of ethics on its own will be of scant use

(Bowie 2009). Accompanying texts, and their institutional context, can be helpful and indeed necessary in their interpretation and implementation (McKerrow 1993). It is often here that key value assumptions are located. Looking at these closely will be especially important in certain contexts: where values are disputed; where values are fundamental and deeply held; where there is rapid technological and societal change occurring. All three apply in the case of AI.

4.2 Professional Codes of Ethics Tend to Have Certain Commonalities

The following is not intended as a full review of the features of professional codes of ethics, but discusses features of particular interest to the question of developing a code of ethics for AI. We need to examine the general rationales for having such codes of ethics or codes of practice in the first place.

4.2.1 Relations Between Professionals, Clients and Others

Gradients of expertise and resources between professionals and others: A code of professional ethics concerns the behaviour and services produced by a professional, who has a certain expertise and who produces something or delivers a service. Thus, the professional has skills and knowledge that the client group typically does not have, producing a gradient of expertise and resources, which then generates a relative vulnerability that gives rise to potential ethical problems that the codes aim to address. In many cases, the professional skill set is accredited, giving prestige to the professional group and presenting barriers to those without the credentials, regardless of their actual level of expertise. The specifics of a particular profession in a particular social context act to shape the resulting codes. Note that their specific professional role gives professionals concomitant additional moral and professional responsibilities; and the opportunities a profession affords also gives opportunities for corruption or unfair use.

Generally, one of the relative vulnerabilities between professionals and others is a general epistemic vulnerability with greater knowledge on the part of the professional, notwithstanding that specific knowledge and practical capabilities on the part of the client might be crucial to the implementation of professional skills. The relative epistemic vulnerability of the client then helps to shape key aspects of professional ethics; for instance: undertakings to assure levels of professional competence, to work only within one's sphere of competence, and to update skills and knowledge appropriately; undertakings of honesty and transparency in dealings with the public and with clients and full disclosure of risks, including taking further advice as needed; undertakings to operate within the law of the appropriate jurisdiction and any relevant local or regional government regulations (which is often simply implied).

The requirements of honesty and transparency will usually involve being able to give an account of actions taken and reasons behind them. An assumption behind this is that individual members of a profession themselves, and the profession as a whole has a significant grasp on its activities and can hence be in adequate control, and at the very least, to insure against unforeseen loss of control.

There will be a working assumption of relative stasis or incremental development in an area, in this sense: that the progress in this area is not outstripping the profession's capacity to understand and control its own area of endeavour. This is of course a matter of degree, since technology and knowledge constantly evolve. But to serve its function, any code of professional ethics has to be capable of addressing significant developments in its area of operation.

Professional codes of ethics are centred on clients but also usually need to refer to the public. The product or service is intended to produce benefit to the clients, and perhaps more widely. There is usually then a concomitant possibility of producing harm, which in the case of some professions can be severe. This harm in particular may affect those other than the clients, hence the need for codes of professional ethics to consider the general public and to make undertakings not to harm (for example, through consideration of the environmental effects of a profession's activities).

4.2.2   Professional Codes of Ethics, Enforcement, and Authority

Codes of ethics ideally outline procedures for reporting problems and violations of codes, which may include protection for whistleblowers and accounts of penalties for proven misconduct. This should draw our attention to the institutional context of professional ethics. Note that there is considerable evidence that, despite professional and legal safeguards, whistleblowers often fear poor treatment and may indeed suffer retaliation (Mesmer-Magnus and Viswesvaran 2005).

The authority and enforcement of codes of ethics may involve professional sanctions, restrictions on membership of professional bodies (which for some professions may make it impossible to practice) and, in worse cases, legal ramifications. Enforcement also occurs through the soft power of the authoritative weight and respect with which the relevant professional body and its codes of ethics are held.

The enforcement of codes may also trade on the relative homogeneity and education of professionals. They have a lot to lose from loss of social standing and income. They have gained a relatively good deal from society, on average. They have been at least to some extent, inculcated into organisations and companies. (This is no guarantor of behaviour, of course. There are many notable examples of spectacular individual failure and institutional corruption. But it forms part of the apparatus of compliance.)

There can be cooperation between different bodies for the enforcement of codes of conduct. For example, concern over the bias in findings of research by pharmaceutical companies by the suppression of negative results has led to moves whereby clinical trials must be openly registered before their start, and academic journals will not publish any trials which are not compliant with this (De Angelis et al. 2004). This may show the effectiveness of outside pressures on professional organisations or companies in helping to change standards of ethics.

4.2.3 Professional Codes of Ethics and Professional Values

There is an assumption of professional value. This relates to a pervading, vital, but sometimes unnoticed background assumption that the product or services of the profession are of general individual and/or society benefit. This assumed value also contributes to the relatively high social standing of members of recognised professions. This assumption is rarely spelled out or argued for in professional codes themselves, but is more implied by the prestige, the training, the professional regard, that surrounds the codes.

Professional practices tend to deal with specific values, arising from a complex of the broad nature of the client group and the nature of the professional services involved. The benefits involved are understood in terms of the particular area of expertise of the profession; avoidance of harms may, of practical and legal necessity, be understood more broadly than the benefits accruing to clients, since they will have to take into account wider consequence. Note, too, that these harms and benefits will tend to be cashed out, not necessarily in terms of a global ethic of human value, but in reference to the particular values of the product or services in question. This will be important in considering codes of ethics in AI.

Linked to this assumption of value, *members of the professions tend to have a relatively high social standing.* Indeed, the very existence of a professional body which produces codes of ethics or conduct also itself helps to contribute to the relatively high status of the professions. Codes may contain undertakings not to bring the profession into disrepute, and undertakings to maintain or improve the social status of the profession. The relatively high social standing also feeds into the soft powers supporting the codes' authority.

4.2.4 Values Underlying Professional Codes of Ethics

There will be explicit values embedded in professional codes of ethics, but also a base of underlying values. The values that lie behind professional codes of ethics will on the whole be values largely shared by the surrounding society, focused towards the particular area of practice of the profession, very often with stricter or additional duties placed on the professionals. As debate and thinking about ethics continues, and as society changes, there may be changes in how these underlying values are articulated and promoted.

However, a fully consistent and agreed set of *underlying* values may be hard to discern. Differences of interpretation and emphasis may mask or reveal deeper differences of opinion, or commonalities, between individuals, groups and communities, and geographical regions, towards these broad underlying values. Even one individual may not have fully consistent understandings of some core value terms: this has been shown for privacy as we saw in Sect. 2.8.3.

4.2.4.1 The Example of Autonomy

Autonomy is not just a core value in contemporary society, not just a core value underlying many codes of professional ethics such as codes of medical ethics, it's of particular concern to us as a key to AI which is developing autonomous systems and machines. It's both a normative value that we aim for in attempts to respect autonomy; and a key notion underpinning our very conception of the moral agent, moral motivation and moral responsibility. It is not just one of our values, it is a presupposition of how we understand our values. It's key, for example, to current understanding of responsibility in warfare, which is challenged by autonomous weaponry (Roff 2013).

Consider: respect for the autonomy of the individual is a core value in codes of medical ethics, expressed in various ways and articulated via concern for issues such as confidentiality and free and informed patient consent. The history of medical ethics over the last century or so can be read in no small way as the history of how patient autonomy has been granted greater and greater emphasis, as opposed to the 'doctor knows best' model (Beauchamp and Childress 2001). At the same time, this then raises questions about the autonomy of medical staff themselves, as seen in debates about the limits of conscientious objection for medics and pharmacists. Such debates are indeed, changing and some would say, undermining, the very idea of the medical profession, and replacing it with a service industry model. There are complex interactions between the expression of value and societal and technological change that it would be very hard to track with complete precision.

There are *philosophical and practical questions* and differences in how exactly the value of autonomy should be understood. Here's one challenge: respect for individual autonomy in clinical medicine may sit in some tension with principles of public health. So we need to understand how to respond when different values that we have clash. The question of the priority of the individual over the group is one of the most central questions of ethics.

There are also large *cultural differences* in how, and to what extent, individual autonomy in medicine is to be valued. A greater emphasis may be placed on community values or social cohesion, for example. Reading journal articles on medical ethics, it's often fairly easy to guess if the authors originated in the USA, or in Northern Europe, by the ways in which autonomy is discussed and ranked alongside other more communitarian or social-oriented values; there are even greater differences visible in discussions of medical ethics from other regions (Padela et al. 2015).

There are *individual differences* in how we value autonomy as well, which may be visible in the work of different moral philosophers, and which also may have strong effects on political affiliations.

Since we are considering the development of technology, note importantly that *scientific findings, technological developments, and brute facts can challenge thinking and action concerning autonomy.* How do we carry on valuing autonomy, for example, in patients with advanced dementia, an increasing problem in advanced societies with aging populations and stresses on social care? (Bridges and Wilkinson 2011). Often, it's advances in science and technology which are presenting us with new, or newly acute issues for autonomy. For instance, the science of genetics challenges simplistic ideas that individuals should have control over 'their' medical information, since genetic information is shared between biologically related individuals (Rhodes 1998). Our views of concepts related to autonomy, such as privacy, individual rights, group rights, and so on, shape our often uncertain and frequently contradictory responses to such developments (Laurie 2001). When we consider the case of AI, the developments of codes of ethics, and assessment of the impact of AI on individuals and societies, we will need to consider such complex interlinking webs.

4.2.4.2 Articulating Values Underlying Professional Codes of Ethics

Providing a definition of underlying values can be surprisingly hard. It's easy to state the goal of medicine is health … or is it the elimination of disease? And how do we even draw a distinction between disease and health—this is much harder than may at first appear, and the philosophy of medicine has long grappled with this question (Boorse 1975).

Definitions of such key terms are not simply there to describe 'reality'. They have a function to perform. We should note that the practice of medicine continues: it's in hard cases that these definitional issues are important, and indeed, they are the stuff of difficult policy debates. Yet, at least in medicine, we are considering a long standing practice; developments in AI may be harder to trace and more disruptive of social practices and values.

A Definition of Health, Extreme Social Change, and Some Thoughts for AI

A widely cited definition of health from Aboriginal Australia states:

Aboriginal health is not just the physical well-being of an individual, but is the social, emotional and cultural well being of the whole community in which each individual is able to achieve their full potential thereby bringing about the total well being of their community. It is a whole-of-life view and includes the cyclical concept of life-death-life.

Health to Aboriginal peoples is a matter of determining all aspects of their life, including control over their physical environment, of dignity, of community self-esteem, and of justice. It is not merely a matter of the provision of doctors, hospitals, medicines or the absence of disease and incapacity. (Houston 1989)

This definition not only includes culture, but justice, and in context, therefore includes consideration of historical events. The health status and life expectancy of Australian Aboriginal peoples is far lower than that of the Australian population as a whole. This very broad definition of health therefore needs to be understood with reference to the devastating impact upon the lives and well being of indigenous Australians by European colonisation of their lands (Boddington and Räisänen 2009).

How is this relevant for AI? There are many who consider that the impact of AI on our lives is not just going to be immense, but also unpredictable. We may not be able to understand what's coming (Vinge 1993).

It struck me that this could perhaps be, in very broad terms, analogous to the unimaginable shifts in life that the indigenous peoples of Australia had thrust upon them. *And so note*, it's precisely this rapid and profound change that's one key motivator for the breadth of the Aboriginal Australian health definition and the reference to culture and history. Likewise, in considering ethical issues arising from the advent of AI, it's likely to be important to look very broadly, and to keep an eye on history and on culture, to consider what is lost and what is changing.

These underlying values may also be up for debate, and here it is particularly pertinent that wider scrutiny may occur. This is especially the case when the actions of a profession have wider social significance. It's vital, too, that academic disciplines taking a lead in ethical discussions in a particular area are self-critical, and avoid domination by particular ideologies or factions. For example, in bioethics, some dominant voices currently are those who take certain utilitarian or libertarian views, and it's been argued forcefully that certain core values, including the value of autonomy and of individual contractual obligations, which are shaping discussion, need urgent examination and critique (Dawson 2010). It's such core values that may also shape discussion of the ethics of AI; we need careful scrutiny and broadly based imaginative thinking.

4.2.4.3   Underlying Professional Values May Be Focused Towards Protecting Individuals

The values underlying a code of professional ethics are shaped by views of primary professional responsibilities. Given the client focus of professions, there may be particular attention to protecting individuals and hence a stress on values which pertain to individuals *qua* individuals, such as privacy, autonomy, and individual property. Codes of professional ethics do however (usually) call attention to the need to protect the public; but a code of professional ethics may well assume that the professional's primary duties are to *benefit* their specific client group, whilst *avoiding harm* to the public. At the same time, a code of ethics may be designed to promote and protect the financial and professional interests of a particular group (Hammersley 2009). Indeed, the very bureaucratic machinery of ethical regulation as a whole might serve the purpose of promoting technological, economic and industrial advancement for a particular group or national region (Dingwall 2008).

Note two points. Firstly, that codes of ethics might then focus on values belonging to a relatively individualistic ethic. And, as valuable and as central the protection of individuals may be, additional values are needed in other contexts. For instance, there are somewhat different considerations operating within clinical ethics compared to public health ethics and compared to medical research.

For, secondly, we need to consider how the relevant services or products potentially affect those individuals who are not clients, and indeed how they affect society as a whole. However, if these questions aren't considered the direct responsibility of individual members of a profession, codes of ethics for that profession may be the wrong place to address these.

A particular problem for AI: Moreover, readily identifiable dangers such as structural collapse or the spread of contagious disease might attract scrutiny, but where technologies are new, rapidly developing, and potentially disruptive or transformative of social relations, as in AI, it will be a complex and often difficult task to ascertain exactly what broader ethical and social issues will arise, and even harder to untangle and trace how a particular technology contributes to these. In such cases, greater scrutiny and careful research to uncover impacts will be helpful, indeed, vital.

4.3   Codes of Ethics and Institutional Backing

A code of ethics is only as good as the institution behind it, and the ethos that operates within that institution. Many a company that has collapsed in the midst of corruption scandals, had inspirational codes of ethics languishing untouched in a golden frame on the CEO's penthouse office wall (McLean and Elkind 2013, 2004). A code of ethics plays only a certain part in the ethical conduct of an institution, and only if it is thoroughly embedded into multiple practices within an organisation can it really have a tangible impact (Bowie 2009).

Broader social and political forces can also undermine the integrity of the best codes of ethics. Take a look at one of the first ever codes of professional ethics for medicine. This code distinguished 'therapeutic' from 'non-therapeutic' research. It included the principles of beneficence and non-maleficence; it was based on an ideal of patient autonomy; it outlined a new legal doctrine of informed consent, which had to be clearly given and based upon appropriate information, with written documentation of consent procedures. There were clear structures of responsibility, and experimentation on the dying was prohibited.

This historically very early and impressive code of medical ethics, 'Guidelines for New Therapy and Human Experimentation', was issued by the Reich Minister of the Interior in Germany in 1931. It was not many years before doctors who were fully aware of such a code were involved in some of the worst atrocities of medical 'experimentation' that human beings have ever done to other human beings (Vollmann and Winau 1996).

4.4   The Context of Codes of Ethics

To understand many codes of ethics fully, we need to examine the institutional background, history and rationales for their production. This context can help us understand what values were addressed, consider how the landscape has changed, and consider who and what has influenced the codes as they currently stand and why. This can help us to think critically about how to amend or develop the codes, and to recall the root values that motivated their development.

For example, codes of medical ethics cannot be fully understood without at least some awareness of the history behind such codes, including the development of medical ethics in the twentieth century since the Nuremberg trials, the development of the Nuremberg code, the Helsinki declaration and its many revisions, and other such developments around the globe (Shuster 1997). Note however, that accounts of the history of ideas and regulations is always complex, and some people contest any simplistic account of ethics regulation as a straightforward attempt to combat abuse, especially as vested interests sometimes may play a part (Dingwall 2008).

> **The Nuremberg Trials: A Baseline of Evil**
> The history of the regulation of medicine cannot be understood without understanding the Nuremberg trials. These addressed the appalling abuse of human beings in medical 'experiments' of profound cruelty. One response to this was to draw up codes of medical ethics to protect the individual to try to ensure that such abuse could never occur again.

(continued)

An observation: the event of the Nuremberg trials must then be seen as pivotal in internationally recognised codes of medical ethics. Why were these codes so readily adopted? Because the abuse of subjects in medical experiments in Nazi Germany was so vile, so inhuman, so degrading, that it is impossible not to consider it an evil. [And note that similar atrocities have been committed elsewhere, for example the inhumane medical experimentation carried out during WW2 by Japanese in Unit 731 (Williams and Wallace 1989)]. It is worth remembering this when the issue of relativism and how to develop and apply codes of ethics cross culturally and internationally is considered. In these and other atrocities, the human person—the most complex creature in the known universe—was treated as a mere subject, a thing.

We can also note a key lesson from Nuremberg. The oft-heard plea, 'I was only following orders' was thrown out as an excuse. The atrocities were far too bad for that. Simply going through the motions, simply following a code, a set of instructions, is not a morality. The individual was charged to stand up against the bureaucratic apparatus of evil, as a few in fact had.

But note that the 'following orders' excuse also reduces the human person as moral agent, to something less than human, a cog in an evil wheel. The attempted denial of humanity was doubly tragic in that it applied both to those who acted, as well as to the profoundly suffering humanity upon whom they acted.

What's this got to do with AI? Since advances in AI are precisely raising questions about the nature of the human agent, and the nature of machine agency; since they present us with potentially profound disruptions to our individual and collective lives; since such changes are happening so fast, it will be as well to recall such fundamental moral starting points as we attempt to think through the ethical questions of AI.

Codes of ethics (and laws and other regulations) have developed in response to catastrophes or scandals, and understanding this can help to understand how codes have grown up as they have. For example, responses to the Tuskegee Syphilis trial have had a big impact upon medical ethics, to name just one of many such instances (Reverby 2012). But as vital as responses to catastrophe have been, developing codes in this way can have pluses and minuses. 'Hard cases make bad law', and responding to something tagged as a 'desperate case' may skew our thinking (Moore 1989). For example, responses to the events such as the thalidomide tragedy made it harder to carry out research on pregnant women, and prevented the use of thalidomide even in patients with scant or zero chance of pregnancy (Benatar and Singer 2000). In AI, one hopes to avoid catastrophe of course, especially if we are talking about existential risk, but we need to consider very carefully how to achieve this.

Extrapolating from Examples, Telling Stories and Gaining Insights

In responding to catastrophe and other bad events, or indeed from examples of good conduct, we are extrapolating from one case to the next. Great care is needed. How cases are described and the context in which they are placed will have a large impact upon how they are interpreted and what lessons are drawn. from them. Tod Chambers' book *The Fiction of Bioethics* shows how by writing and re-writing cases, different interpretations and conclusions may be drawn (Chambers 1999). It's often common, and understandable, to focus on the most graphic and extreme cases, but this can demonstrably skew resulting policy (Boddington and Hogben 2006).

Note, too, that the way a case is described might block or permit our own moral insights. King David had an adulterous affair with Bathsheba, and, after she became pregnant, deliberately sent her husband, Uriah the Hittite, to his death in battle. The prophet Nathan described a case of a rich man taking a poor man's sheep; King David denounced the actions of the rich man in taking what little the poor man had. By packaging the essentials of King David's heinous acts in parable form, Nathan presented David with the uncomfortable truth: Thou Art the Man (2 Samuel 12) (Butler 1827; MacNaughton 1988).

Note that often it is a third party coming from an outside perspective who is best placed to do this; and someone who is prepared to speak truth to power. In order to keep an outside perspective on one's moral values, Philip Zimbardo argues for the ethical necessity of belonging to more than one social or peer group (Zimbardo 2008).

Codes and regulations may be developed in anticipation of possible problems. For example, the EPSRC Principles of Robotics may be seen as an attempt to avert the kind of public backlash that was seen in the UK over GM crops (Bryson 2012). Hence, such background issues are again important to understand in considering the purpose and final shape of any codes or regulations. The recent government documents such the House of Commons Science and Technology Report (Robotics and Artificial Intelligence 2016), the European Union's Committee on Legal Affairs Report on robotics (European Civil Law Rules in Robotics 2016), and a report by the Obama Whitehouse (Executive Office of the President 2016), are attempts to anticipate particular problems within particular political landscapes.

Codes of ethics may develop in response to other codes of ethics. This may or may not be appropriate. For example, codes of ethics for social science researchers have been historically modelled closely on codes of ethics for medical research. But the risks involved in social science research tend to be of a quite different kind, and of a different degree. Moreover, social science research methodologies may differ greatly from those of medical research (Atkinson 2009). The regulation of social science research has suffered in many respects from being shoehorned into a medical model. We need to think carefully about how AI, in its many different forms, needs to be regulated, rather than simply tinkering with what we already have, and rather than assuming that the same model of codes and regulations will do for all forms of AI.

Codes of ethics may be influenced by certain powerful groups or individuals. This can be useful, but there are drawbacks. Those working on public engagement have long recognised that there are multiple 'publics', and that different interest groups among the public may have quite different agendas. Some of these may be avidly pro-science (Novas and Rose 2000). Others may have very different views (Plows and Boddington 2006). Organised groups may be influential, perhaps unduly so; and membership may be skewed, with those who don't join groups likely to have different opinions. Many patient groups may act as lobbyists, receiving funding from the pharmaceutical industry (Herxheimer 2003). Some codes or sets of principle themselves are of course produced by powerful groups, such as prominent members of a profession, for example the 1975 Asilomar Conference Recombinant DNA Molecules (Berg et al. 1975). One take on the groups producing such statements is that these are the ones who know what they are talking about. Another take is, yes, sure, but others need to have a say as well, and are likely to have very different interests. Yet another consideration is how such groups are selected, or self-selected, for influence and persuasion.

Codes of ethics of professional bodies also often have a wider national or international context. For example, codes of medical ethics for different countries exist in the wider context of the policies of the World Health Organisation. Codes of medical ethics are closely linked to the development of medical law in the relevant jurisdiction; and the development of medical law in separate jurisdictions is itself often influenced by developments and cases in other jurisdictions. Much research takes place in a global context. For example, much research in genomics of necessity needs to study different population groups of humans in order to conduct scientifically robust research. Complex ethical considerations of how to marry global standards with local sensitivities may be needed (HapMap 2004).

Codes of ethics also develop in relation to certain cultural contexts, and these may influence them in ways which are hard to discern, especially if we are also embedded within that context. The development of law, regulation and practice within different geographical areas in itself helps to shape this cultural context. For example, laws regarding the protection of privacy in the use of personal data within the EU are currently more stringent than in the US; this helps to shape debate and opinion, but it's not clear that this difference in emphasis could have been predicted in advance. This again helps to illustrate how complex, interwoven, and perhaps unpredictable, are such developments in values.

Clues to influential cultural context may be found in literary devices such as the rhetoric used in surrounding text, and allusions made. In the context of technology in general, and AI in particular, reference to science fiction and to various stories regarding robots, computers, and out-of-control creations is frequently made. These may be instructive of underlying beliefs and values.

Science Fiction and Myth in Policy Making for Robotics
The European Parliament's Legal Affairs Commission have published a study on 'European Civil Law Rules in Robotics' (Directorate General for Internal Policies 2016). It illustrates how reference to myth, legend, science fiction and popular culture are routinely referred to in policy and ethics discussions regarding AI in general and robotics in particular. But note how the rhetorical reference to such stories can help shape the thinking that then goes to frame how the surrounding policy is read and understood. Here, 'Western' responses to robots are contrasted with those of the 'Far East':

1 Western fear of the robot
The common cultural heritage which feeds the Western collective conscience could mean that the idea of the "smart robot" prompts a negative reaction, hampering the development of the robotics industry. The influence that ancient Greek or Hebrew tales, particularly the myth of Golem, have had on society must not be underestimated. The romantic works of the nineteenth and twentieth centuries have often reworked these tales in order to illustrate the risks involved should humanity lose control over its own creations. Today, western fear of creations, in its more modern form projected against robots and artificial intelligence, could be exacerbated by a lack of understanding among European citizens and even fuelled by some media outlets.

This fear of robots is not felt in the Far East. After the Second World War, Japan saw the birth of Astro Boy, a manga series featuring a robotic creature, which instilled society with a very positive image of robots. Furthermore, according to the Japanese Shintoist vision of robots, they, like everything else, have a soul. Unlike in the West, robots are not seen as dangerous creations and naturally belong among humans. That is why South Korea, for example, thought very early on about developing legal and ethical considerations on robots, ultimately enshrining the "smart robot" in a law, amended most recently in 2016, entitled "Intelligent robots development and distribution promotion act". This defines the smart robot as a mechanical device which perceives its external environment, evaluates situations and moves by itself (Article 2(1)). The motion for a resolution [calling for 'the immediate creation of a legislative instrument governing robotics and artificial intelligence to anticipate scientific developments over a medium term', p 8] is therefore rooted in a similar scientific context. (p 10)

Here, the West is seen as having a negative attitude of fear towards robotics, and the document itself then expresses a fear of its own, that this may 'hamper the development of the robotics industry'. But note how Western attitudes are presented as emanating from tales and myths, framed only as 'ancient tales' and 'romantic works'. However, the positively presented Far Eastern attitudes are presented with a more solid underlying metaphysics or ideology—in Japan, that of Shintoism. This contrast then automatically

(continued)

frames the Western responses as shallower; subliminally, it's as if the West just scared itself with 'spooky stories'. The motion for a resolution proposed in this policy document is then nested in the positively framed Korean response to legal instruments regarding robots.

However, there are significant currents of Western thought and writings which could be used to explicate a response of fear to robots, which could provide a well-articulated and long established underlying basis of thought. For example, the influential creation story of Genesis presents a creator God as making autonomous beings—us—who quickly behave atrociously and disobey their Maker. Cain murdered his own brother, Abel, in a jealous rage; and on the story goes. So, it seems no mere coincidence that fears about the behaviour of autonomous robots would be strong in Western literature. The control problem of autonomous agents is also precisely a major concern of current attempts to address ethical issues in AI.

Note, too, that the EU report in fact goes on to reiterate that there is reason for Western concerns, 'now that the object of fear is no longer trapped in myths or fiction, but rooted in reality' (p 13) and cites Bill Gates, Elon Musk, Stephen Hawking, and Bill Joy as issuing warnings regarding autonomous AI. Note, however, that this subtly juxtaposes the recent reasoned warnings of scientific experts against the ancient myth-and-story driven fears of the populace, which perhaps by serendipity happen to coincide. This perhaps is not a useful way for a policy document to frame the concerns of 'experts' versus 'the public'. This is especially true given the way that AI does in fact raise questions pertinent to the very foundations of our morality and of our view of ourselves. The document then subtly priorities the expressed concerns of technological 'experts' while diminishing the concerns of 'the Western collective conscience'.

4.5   Can Codes of Ethics Make the Situation Worse? Yes

We've seen that codes of ethics need a strong institutional backing to function effectively. But codes of ethics can actually make matters worse.

Separation of ethics from 'life': The very idea of parcelling ethics into a formal 'code' can be dangerous, if it leads to the attitude that ethics itself is just some separate part of life and of activities. Such a risk exists if the code is presented as a set of instructions to user. 'Perceptions that a code presents the voice of an external authority frequently go along with a defensive and punitive institutional ethos that suggests to code users that it is necessary to lie low and keep out of trouble in order to avoid threats of criticism, negative judgement and punishment' (Bowden and Surma 2003) p 26.

'Can't someone else do it?' Homer Simpson once ran for Sanitation Commissioner of Springfield under this banner (Trash of the Titans 1998). It didn't go so

well. The existence of a code of ethics would be problematic if it encouraged the idea that it was *somebody else's job* to 'do the ethics'; although there can be good reasons to ensure that specific nominated individuals are assigned responsibility for certain issues, as a check against the *diffusion of responsibility* within organisations and looser groups. The appointment of a role such as a 'Chief Values Officer' *might* present such a danger, depending on how the role was implemented. The ways in which responsibility is avoided by individuals and diffused within institutions has been discussed in relation to the very large and often geographically dispersed groups of researchers that may be working on a project (Caulfield et al. 2008). Work in social psychology has turned up some valuable lessons in how easy it is to create the conditions which allow for diffusion of responsibility to occur (Zimbardo 2008). It would be very worthwhile for those considering the effectiveness of codes of ethics for AI, which may be developed by very large groups of people working on different aspects, to contemplate these problems.

And codes and regulations may encourage 'work to rule'—to work up to the regulation, up to the code, and no further; to the letter of the code, not the spirit. This may be especially problematic in some areas, such as those pertaining to safety. Well known examples of operating to the letter, not the spirit, include 'shopping around' for a tame ethics review board, or operating in countries where the standards are not so tight (Gulhati 2005).

So, a code of conduct might produce a 'tickbox' culture of ethical complacency where filling in and complying with paperwork becomes an end in itself, and the goal of ethical compliance is focused on too narrowly, and for the sake of reward or of avoiding penalties. For example, in some situations this may apply to the practice of obtaining informed consent to medical treatment and/or medical research, where the staff have the task of 'consenting the patient' (Jones 1999). Worse, such a mentality can encourage the very behaviour it was intended to discourage (Bazerman and Tenbrunsel 2011; Adams and Balfour 2014).

Indeed, the existence of a code of ethics or other systems of ethical guidance may give rise to ethical display behaviour or 'virtue signalling' (Bartholmew 2015). Easily signed declarations of ethical intent may have no impact and may entice signers to overinflate their self-assessed moral character (Bazerman and Tenbrunsel 2011). It's always worth remembering that in Stanley Milgram's classic Obedience to Authority experiments, where subjects were led to believe they were involved in an experiment on learning and that they were delivering electric shocks to 'learners' (actually stooges), Milgram found that expressing moral doubts enabled subjects to retain a self-concept as a 'good' person, and actually made it easier for them to continue to administer 'shocks' to the stooge (Milgram 1974).

Let me repeat that: *expressing a moral sentiment may in some circumstances decrease the likelihood of behaviour that follows one's conscience.*

There are indeed very hard questions about how to translate institutional ethical policies into practice. For instance, recent work on the 'paradox of meritocracy' shows that institutions which consciously flag meritocracy may in fact show greater bias towards men over women than those which do not (Castilla and Benard 2010), and ethics and HR policies which mandate the currently fashionable implicit bias training must face mounting evidence that such training may even make the situation worse (Duguid and Thomas-Hunt 2015). Any code of ethics which wishes to encourage such good behaviour thus needs to take careful heed of research and developments on the question of how best to bring about such changes.

The flip side of this is that where codes of ethics are *unduly* restrictive, there may be some justification in giving them short shrift. A code of ethics might worsen a situation by tying the hands of professionals whilst those outside the profession can carry on a practice with impunity. For example, restricting research into a particular area because of its dangers might mean reduced capacity to counter any dangers in that area from competent outsiders to a profession.

Ethical Arms Races and Being 'Too Good'
In the children's book, *Super Duper Jezebel*, the main character is a goody-two-shoes little girl who never breaks any rules (Ross 1988). One day a crocodile enters the school playground; refusing to break the rule against running at school, Jezebel alone comes to a sticky end. In a related vein, Hilaire Belloc's *Cautionary Tale* of the truly nauseatingly good child, Charles Augustus Fortesque, paints a comic yet starkly unattractive picture of an entire, unimaginative life lived according to blind conventional attachment to social rules (Belloc 1907).

There is a particular problem with slavish conformity to ethical rules in an unruly world full of 'bad guys'—suppose we render ourselves vulnerable to calamity by ideals of ethical purity that don't equip us to fight dirty if push comes to shove? It's also one thing to sacrifice yourself to a moral ideal, another entirely to sacrifice *others* to your ethical ideals.

Unreflective conformity to existing moral and social convention is particularly problematic where there is a need to address flaws in the existing conventions. This does not necessarily mean that we want everybody to be always questioning existing convention. But we need a dialogue with those who are raising concerns and pointing to shortcomings. There's a particular problem where the wish to conform to otherwise laudable rules makes us powerless in the face of those who flout the rules. As we've seen, OpenAI in fact specifically aims to combat such a problems by producing open source AI, in the hope that this will help to undermine the potential dangers from those creating malicious code (OpenAI).

The very real problem of how to avoid an arms race of autonomous weapons is mentioned here but it would be foolish in a book of this generality and length to attempt to do anything more than point out its difficulty (Roff 2014).

And while we're on this topic: experiments such as Milgram's, and others such as the Stanford Prison Experiment, produced insights into human moral

(continued)

behaviour, yet, therefore perhaps inevitably, had dismaying effects upon the subjects. Philip Zimbardo's Stanford Prison Experiment randomly assigned students, who had all been screened for good mental health, to the role of either 'prisoner' or 'guard'. Many of the 'guards' then meted out harsh treatment to the 'prisoners'. Tellingly, the experiment was called off early only after Christina Maslach, who was not directly involved, observed what was happening and protested. Zimbardo had the sense to marry her (Zimbardo 2008). We've seen how extrapolating from catastrophe can have problematic results as well as good. Such experiments have fed into ethical regulation of social science research which now makes it virtually impossible that such experiments could today get ethics clearance. Yet, these experiments helped to give us valuable insights into unethical behaviour. This is a particularly perplexing conundrum.

Keep calm, dear: There is also a danger that a code of ethics might actually be serving the purpose of calming public anxiety, without actually managing to make an iota of difference to the substance of warranted public concerns. Ethical regulation of new technology might serve to placate concerned groups and individuals and the very existence of regulation around a new technology or practice might make the unpalatable seem more palatable (Bryson 2012; Dingwall 2008).

'Administrative Evil'

Work such as the book *Unmasking Administrative Evil*, which looks closely at the root sources of many technological catastrophes and institutional failings, is extremely pertinent to the consideration of developing codes of ethics for AI (Adams and Balfour 2014).

The failure to apply policy correctly can be a big problem, especially where this failure emanates from pervasive institutional and leadership failings.

But in some cases, it is the very *application of policy* that can inadvertently give rise to deleterious effects, sometimes effects precisely opposing the intent of the policy.

This concerns a 'technical rationality' and how value issues can get lost within large systems. The dangers increase the greater the efficiency of the system and the greater its automation and distance from (uncorrupted) human affective response.

However, it is perhaps not beyond the wit of those studying autonomous systems to consider how to combat such potential for administrative evil. It is a recommendation that such a possibility is studied closely by those drawing up codes of ethics for AI within organisations and systems.

Regulations and boiling frogs: And, since codes of ethics are developed over time, what's seen as problematic is gradually changed or eroded, so that practices can be introduced little by little. This is often known as the 'boiling frogs' problem, although the jury's still out on whether actual frogs do this. (Don't try it out. You'll never get ethics clearance for the experiment.) For example, as recently as 30 years ago, the consensus in the UK at least was that we would never even consider sex selection of human embryos for social reasons. Now, whether for good or for ill, precisely this question has been considered (Holm 2004). Is this progress? Or the reverse? Or is this simply change, and nothing more? Importantly, in the development of regulations in emerging fields, one might expect that changes would be made as our understanding of the technologies changed. But it's common to draw 'lines in the sand' that will 'never be crossed'. Yet, such 'lines in the sand' often do then get put up for further consideration. Is there then no such thing as a 'line in the sand'? Often we are nudged into considering blurring policy lines by the directions of the technology.

Now, this might on the one hand be a genuine acceptance of the new; perhaps with a realisation that it hasn't brought the feared changes, or with a reappraisal of what counts as a plus or a minus. But it could also be step by step introduction of changes which in the end add up to a large change which would never have been accepted, had it been introduced all at once. After all, that is precisely how Milgram got decent individuals to deliver electric 'shocks' to strangers—little by little (Milgram 1974). This will be particularly hard to assess where rapidly developing technology that is embedded in our lives in multiple ways is having a large impact upon how we live.

Oh wait. That's happening with AI.

Chapter 5
How AI Challenges Professional Ethics

Abstract Having considered those aspects of professional practice which underpin
the need for professional codes of conduct, and the nature of such codes, we turn
now to consider how AI presents particular challenges for developing professional
codes of ethics. Although many working in AI may be members of a professional
body, work in AI may be carried out by those outside of any formal organisational
setting. In addition, in AI, resources of money, technological capacity, and capacity
over the dissemination of information, may be concentrated into certain hands. At
the same time, the control problem in AI undermines the expertise gradient that in
most professions gives them power and authority. This means that in AI, there is a
particular problem with professional vulnerability in relation to their own products.
Codes of professional ethics generally deal with two matters: the behaviour of
professionals, and the impact of their products or services on clients and on the
wider public. In AI, we have a third, additional, layer of complexity that must be
addressed: the behaviour of machines.

AI challenges the standard model of a code of professional ethics on many of the
major features outlined in Chap. 3.

5.1 AI Professional Organisations and Companies, and the Nature of Its Development and Production

The production conditions were briefly considered in Sect. 3.4. I now consider in
more detail these issues in relation to professional codes of conduct.

Workers outside professional bodies: Although many working in AI may be
members of professional organisations, such as the IEEE, many are not, and need
not be. But AI can be developed by people outside of any professional or formalised
credentials. We need then to recognise this fact.

Independent and perhaps isolated work possible: To run a lab developing some
AI applications, such as autonomous vehicles, requires large resources. But the very
varied nature of AI means that some self-taught, technically competent person, or a

© Springer International Publishing AG 2017   59
P. Boddington, *Towards a Code of Ethics for Artificial Intelligence*,
Artificial Intelligence: Foundations, Theory, and Algorithms,
DOI 10.1007/978-3-319-60648-4_5

few members of a small scale start up, could be sitting in their mother's basement right now dreaming up all sorts of powerful AI. No accreditation is needed. This person might be an evil maniac, or, possibly worse, a reckless well-wisher with a Messiah complex. In fact, the capacity of the 'lone wolf' computer hackers of modern mythology to disrupt our technology-dependent lives is possibly one of the underlying drivers of fears concerning AI. Combatting any ethical problems with such 'wild' AI is one of the major challenges.

Concentration of both finances and resource-share into a few hands: The sheer amount of money that some organisations or even individuals have to work on AI (and related technology) is also a feature of concern. There is little doubt that this brings great power to develop AI. The increasing steepness of the gradient between the world's richest individuals and the rest of the population means that sheer financial power may be coupled with the corralling of resources into a few hands which has been a feature of the development of much IT (Piketty et al. 2014). This further steepens the power gradient of various key actors in AI in relation to others.

Resource concentration is not just limited to finances and technological capacity. For example, those with control over online resources have immense power; one major area where AI is also involved concerns the access to, and quality of, information on the internet, and this information, including personal data, can be used and abused. On March 12th 2017, Tim Berners-Lee, founder of the World Wide Web, published a letter voicing various concerns which are also shared by many others (Berners-Lee 2017). Such concerns include the capture of personal information and its use, via algorithms, to target political advertising to certain audiences, as well as what's recently come to be labelled as 'fake news'.

Despite having warned readers of the danger of hype in AI, it is nonetheless important to note the reach and complexity of such problems. In order to advance into the future of AI to the benefit of humanity, it's critical that there is widespread understanding and debate. But the spread of (mis)information and influences on opinion, together with the capacity of social media to silo people into like-minded groups which often seem to be ranged in stark hostility to other groups, is a worrying problem which is both a product of complex socio-technological developments that include the application of AI, and a limitation on precisely the clear thinking and public debate needed in formulating and implementing ethical AI and codes of ethics (Pariser 2011).

Very mixed motivations and attitudes may enter the play: In the current AI landscape there are those with aims somewhat at odds with each other. Just as there are many warning about the dangers of AI and fearing the advent of the Singularity, and suggesting that we maybe need to slow down a bit while we have a bit of a think, there are those working just as avidly to bring this about (IEEE Spectrum 2011).

AI covers a very wide remit of developments and applications, so there will need to be many specific codes of ethics for different AI applications and developments. As we've seen, professional codes of ethics are cashed out in the relatively specific values associated with the particular products or services of the profession in question. A general code of ethics for AI would have to be at a very broad level of specificity, with more detailed codes needed to translate general principles into workable action in specific domains.

The values underlying codes of professional ethics: Note that our discussion so far indicates that AI raises questions that reach not just to the values embedded in codes of ethics, but to the fundamental groundwork of ethical systems to the heart of what it means to be human, what it means to have agency in the world, and what activities individual humans—and society in general—wish to value. *We can conclude then that in its typical use, AI then has the potential to disrupt our thinking about ourselves, our natures, our capabilities, and our place in society and in the world in ways that may cause major upset to the underlying ethical thinking behind codes of ethics.*

In comparison to the relatively regulated, and credentialised older professions, these features put together make AI a veritable ethics regulation Wild West; moreover, a Wild West where we're not too sure who should be wearing the sheriff's badge. We will then need to think carefully about how to proceed, what needs to be included in any codes, how codes are to be developed and embedded in institutions, and whether codes of professional ethics are going to be on their own adequate.

5.2   Gradients of Professional Power and Vulnerability in AI

Recall the discussion of the assumption of professional power which creates a gradient of expertise relative to those outside the profession, and which underpins and shapes any codes of ethics. Now one of the major issues flagged in relation to AI is precisely the fear that even the professionals might be relatively vulnerable in relation to AI, especially given its complexity, its power, and the speed and range of its development (Bostrom 2014). Fears about encroaching loss of professionals' control over the fruits of their labours do not make it unique *per se*. See for example, worries about biotechnology 'escaping' the lab' (Koepsell 2009). But the extent of these fears with respect to AI are *at least as great* as that of any other technology, especially as AI has such wide potential applications.

It will be helpful to tease apart some different sources of this disruption to the standard gradient of vulnerability assumed in professional codes of ethics. Some relate to knowledge and understanding, leading ultimately to questions of control.

The nature of much AI gives rise to *epistemic vulnerabilities* for the professionals developing it. For instance, an epistemic 'black box' vulnerability arises from the development of machine learning which means that its creators may have only scant knowledge of what is going on inside the black box, or even none. This produces problems with prediction, explanation and justification; recall that justification of actions to appropriate others is a key element of ethics.

Note that this is an 'in principle' epistemic difficulty. Compare medicine. Doctors often don't understand fully how a treatment works, and can predict deleterious effects only imperfectly; because of this, there is a pretty robust system of regulations and accountabilities to protect patients, and meanwhile, medical knowledge steadily increases. In contrast, AI might make unpredictable decisions that we cannot *even in principle* explain. AI needs to produce such a parallel system of safety checks; but in at least some cases, will be doing so against a barrier of epistemic uncertainty. There are differences of opinion about the technical feasibility of achieving transparency, with suggestions that machines could be programmed to explain to humans what they are doing and why, but no consensus on whether this can in all cases be achieved.

The epistemic problem steps up a few notches when we consider the prospect of superintelligence. If this is realised, there won't just be problems like not being able to explain exactly why an autonomous machine programmed to perform certain functions performed them in certain ways. We might not have any earthly clue what a superintelligence is even up to, or why (Bostrom 2014).

So, in worrying about epistemic limitations, there are fears that AI may *develop in ways that surpass human control*, so that it might literally outsmart its creators, and it will then start to control us; that we won't fully understand what we've unleashed. This fear may apply in many areas of AI, and not just to AGI, artificial general intelligence or superintelligence. An aspect of the fears about loss of control is the worry about how AI may be shaping our world in ways which may be hard to detect, poorly understood, and hard to control. How do the algorithms which shape social media shape our thinking and our social relations in unpredictable and possibly unappealing, even dangerous, ways? What implications are there for social relationships, for brain development even, for how we make judgements about what to believe and what not to believe?

Deepening fears about loss of control concern *what might happen to AI 'in the wild'* once released from the immediate control of its creators. A particular fear is that an unstoppable or unpredictable set of natural reactions might occur, resulting in self-replicating and uncontrollable AI. AI is not entirely unique here. Worries that products might be unstoppably self-replicating occur in relation to genetically modified organisms, nanotechnology and the scare about 'grey goo' (Clarke 2005; Oliver 2003). As with these other areas, one challenge for regulation is to ascertain a reasonable response to unknown and possibly overplayed, or underplayed, risks.

As with many other areas of technology and of knowledge, the growth of AI can lead to *increasing specialisation* so that individual professionals may have good levels of knowledge in relation only to increasingly isolated areas of expertise, producing a relative vulnerability for individual professionals and for the profession as a whole, and presenting challenges of communication. It should be noted however that analysis of networks and systems is a key aspect of AI. It could be very useful to draw on such expertise within the AI community in the development of ethical guidance.

The Sorcerer's Apprentice

A poem written by Goethe tells of a sorcerer who leaves his apprentice alone and unsupervised (von Goethe 1878). The apprentice decides to cast a spell to make the broom fetch water. As a result of problems in specifying the task and in controlling the broom, there were uncontrolled and disastrous results.

Note that the moral of this tale seems to imply that a fully qualified Sorcerer, or a fully trained or supervised Apprentice, would not have produced such mayhem. In relation to AI, at least in certain forms, this might be optimistic.

This tale also alerts us to a serious difficulty with the idea that human control over AI will be sufficient:

Humans can be really foolish. They can fail to think through consequences. And they may have an unrealistic idea of their own agency, and rush to take control as soon as they can.

Note two elements which can be teased apart in Goethe's poem: incompetence and lack of supervision. Many argue that AI should always be under ultimate human supervision. But humans themselves also need supervision. Research in social psychology finds that, unsurprisingly, behaviour tends to worsen on the 'night shift', when other people, including those with more seniority or in positions of management, are not around (Zimbardo 2008).

5.3   A Third Layer of Complexity in Codes of Professional Ethics for AI: The Behaviour of Machines

Codes of professional ethics typically deal with two elements: the behaviour of the professionals themselves and the impact of their products or services. However, in dealing with autonomous machines and autonomous systems, the control problem means that a third element must be introduced—the autonomous and often opaque behaviour of those machines or systems themselves. Note how this then necessitates another layer of complexity in ethical codes.

This means that codes of professional ethics for AI must take a significant departure from other codes of professional ethics.

Dangers of Robotics: Does Europe Need an Agency for Robotics and Artificial Intelligence?

Are professionals working in robotics able to spot any potential dangers of their work, or is there a need for oversight of such work at some distance?

The EU document on European Civil Law on Robotics (Directorate-General for Internal Policies 2016) suggests that where autonomous robots

(continued)

pose a danger to humanity, research needs to be regulated if not prohibited, and suggests that a future European Agency for Robotics and Artificial Intelligence could take responsibility for identifying areas of potential danger. It suggests that self-replicating nanorobots might be one such area, and that there would be a need for an external control of such research activities. (p. 13)

The wider public interest in the potential dangers of robotics and AI research, does legitimately appear to be one respect in which some external scrutiny would be justified. This would mirror existing situations, for example with regard to issues pertaining to public health, or indeed any other issues with wider public interest, such as planning regulations in architecture and pollution control laws.

A body with oversight of various groups working on robotics might be in a position to determine wider threats and trends which might not be so visible closer to the ground. Much would depend upon the powers that such a body has in its constitution and membership.

5.4   The Authority of Any Resulting Codes

As we've seen, professional codes of ethics generally presume a set of professionals operating within a social and legal framework where it is clear who is and who is not a professional, and where there are professional bodies that can thus make statements and potentially act as enforcers of codes. 'Soft' regulations, such as professional codes of conduct, only have teeth if they have some sufficiently weighty institutional backing and a reasonable consensus within the relevant community. It's important to note, too, that ascertaining a consensus is no easy matter; those within the umbrella of a professional organisation will at least have some institutional structure within which a methodology for producing and ascertaining consensus can be implemented. Within more loosely organised areas of endeavour, such as in AI, consensus building is much more difficult. Codes also, importantly, require an adequate degree of backing from users of their products and services and from the general public—a degree of backing which is necessary even if there is the hard force of law behind it. But this then gives a particular issue for codes of ethics in AI.

So, what happens to the institutional and social backing, the soft power, necessary for the proper implementation of codes of ethics, in the light of the problems of relative professional vulnerability in AI? The work currently underway by the IEEE in their Global Initiative for Ethical Considerations in Artificial Intelligence and Autonomous Systems has the advantage of the backing of this large and prestigious organisation, which currently has over 400,000 members worldwide and is the largest professional body of its kind.

5.5 Conclusions

Features of AI and its development are pushing in opposing directions in ways which increase dramatically the scale of any task facing those developing codes of professional ethics. We have increasing power gradient for (some) actors in AI over others, in economic power and in organisational resources and capabilities, with the concentration of capacity in certain hands and in certain regions of the world.

Yet, at the same time, those creating and using AI themselves have increasing vulnerability of knowledge, understanding and control towards their creations, representing a weakening of professional power. This is the very reverse of what we would hope for creating the conditions for the development of beneficial and ethical AI.

Even on a model of professional codes of ethics which assumes that the professionals have a high level of control over their products and services, there may be, and surely should be, opportunity for clients and interested members of the public to have some feedback and input into those codes, especially where significant societal and technological changes are occurring. In the case of AI, the fact that professionals are themselves having directly to grapple with how to control the behaviour of machines surely gives end users, and the general public, considerable reason to be kept informed at the least, and actively involved at best. This strongly suggests a very pressing need for significant input and oversight from others in AI, a need which I suggest outstrips even the need for input from outsiders and stakeholders in codes of ethics for other areas.

Moreover, strategies such as the attempt to build ethical behaviour in machines requires considerable thought about the nature of ethics, of moral autonomy, of choice, of agency, of responsibility; such work is too important to leave to AI professionals, and will require a wide range of input. This is especially the case given our discussion above about the profound questions that AI potentially raises about ethics.

In considering the development of codes of ethics for AI, some of the challenges facing us are general challenges about the rapid or unpredictable development of technology, especially where that technology is deeply embedded in social and economic practices. We now turn to the question of producing ethical guidance and codes in situations of fast change.

Chapter 6
Developing Codes of Ethics Amidst Fast Technological Change

Abstract As with other technologies, one of the tasks facing those concerned with AI is how to develop codes of ethics in the face of rapid and perhaps unpredictable technological change. We need to consider carefully what methodology we use in considering ethical questions: abstract principles and accounts of virtue based on the past may or may not be appropriate, and a consequentialism which assumes the assessment of harms and benefit may likewise be of scant use where social and technological change makes the identification of benefit difficult. Close attention to context and complexity will be needed. Technological change and social change are closely intertwined, and may impact on many of our key value concepts, including those which may not at first sight be carriers of social and cultural value; in considering AI, we must watch out for such effects. The complexity of the question of technological changes brought by AI is illustrated by briefly considering the impact of AI on employment and by asking questions about the value and meaning of work. Technological changes are likely to have global impact in at least some cases, and the implications of this for how we think about universal values such as human rights, for relativism, and for cultural diversity are briefly considered. Lastly, as one solution to the complex task of developing ethical guidance for AI, a proposal is to ensure diverse participation in discussions; what is meant by 'diversity' in this context, and why it is essential especially for AI, is discussed with reference to some recent research findings.

6.1 Social, Cultural and Technological Change and Ethics

There are many questions that need to be asked in considering how to develop codes of ethics in rapidly developing technologies. How can ethical regulation keep up where technologies are developing so fast? Contrariwise, how can we ensure that ethical regulation does not jump the gun and produce inappropriate guidance, or stifle beneficial developments? How can the public be kept informed and involved in debates about rapidly emerging technologies? How can we even measure the impact of technologies, and the impact of any codes of ethics, under circumstances of rapid and wide-ranging change? Technology advances alongside and interactively with social changes, changes which include how technology is itself viewed.

© Springer International Publishing AG 2017

P. Boddington, *Towards a Code of Ethics for Artificial Intelligence*,
Artificial Intelligence: Foundations, Theory, and Algorithms,
DOI 10.1007/978-3-319-60648-4_6

This complicates our ethical responses to its development. So, given rapid and uncertain change, how can we progress in attempting to develop codes of ethics for AI?

6.1.1   Methodology and Moral Theory in Times of Change

One simple account of tackling ethical issues is that you take a general moral theory and apply it to a particular case. This model has been frequently critiqued by philosophers (MacIntyre 1984; Archard and Lippert-Rasmussen 2013), and in any case, it gives an unhelpful account of what our task is. If you were completely committed to a particular moral theory, and if you were starting off from scratch in some alien world or totally novel case, and if there was no one else around whose opinions on ethics needed to be taken into account, you might use such an approach. But that is not our task in an area like AI, which is rapidly developing in a world where change is constantly happening; which is developing in ways closely inter-woven with other changes; and which is in various complex ways embedded in our lives.

We need to think about the ethical questions that AI presents. And we need to think about how we might produce codes of ethics or regulations which address these ethical questions, practically and to some beneficial effect. We need to think carefully about methodology. We are not starting from scratch. Our task is to look at how we are thinking about things right now; to think about what we like and what we don't like, to understand how we got to where we are now, what we want to take on our onward journey, what might be left behind. We need to then look back to the past at the very same time as we are looking to the future.

We might then draw on consequentialist thinking, which tends to be the sort of ethical thinking above all others which is free from context and culture, and which then might be useful for social reform, indeed as it historically has been. We might also draw upon ideas of abstract principle, which transcends particularities, seeing carefully how these might or might not apply. We might draw upon virtue ethics, but this will of necessity have to ask if the virtues of the old world are suited to the new world that we are creating. We are rewriting the world in which we live, as we live in it, and our ideas about how to live in it are shaped partly by the very changes we are making.

For example, drawing on a consequentialism which gives happiness and the avoidance of unhappiness as a basis for ethical decision-making might be tempting for its simplicity, for it gives us some seemingly tangible measure of success: does this use of technology produce the most happiness, or not? But a great part of its appeal lies in offering a relatively simple account what outcomes it measures; insofar as it does this, it tends to strip away our attention to these very issues of how we are located in the world, the issues pertaining to our accounts of human and moral agency, and social relationships, which as I've argued, are profound ques-tions which tend to be characteristic of AI. Moreover, a consequentialism such as utilitarianism tends to work best when we are dealing with things we can definitely identify as pointless misery. The great eighteenth and nineteenth century utilitarian social reformers were at their most convincing when they used utilitarianism as a corrective to manifestly avoidable suffering. Using happiness and unhappiness as a tool for ethical decision-making may perhaps work in a relatively stable state, and when we have relatively clear ideas about what is and what is not going to make us happy or unhappy. But where significant change is at issue, any difficulties in its application are compounded.

Keeping Oversight in Pace with Changes in Technology: The Example of Adaptive Licensing
The release of new pharmaceuticals onto the market is tightly regulated, resulting in a development process that can take many years. Often, long before a drug is licensed for use, early clinical trials show some promise, yet the drug remains unavailable to patients not enrolled in the trial. Adaptive Licensing is a way to release drugs to patients earlier than previously possible, with their effectiveness and any possible side effects closely monitored in real time. Technological advances including in the feedback and the analysis of data have in fact rendered this possible, demonstrating that technology is not just a headache for regulation, but can be helpful. Perhaps such a model of tracking use and impact of technology could be one for some forms of AI? Steps in medicine to formulate regimes for Adaptive Licensing of pharmaceuticals to facilitate their use earlier in the development process also have some parallels with the kind of concerns about release of AI into the wild (Oye et al. 2013).

6.1.2   Social and Cultural Change Is Perennial

If ethics is about how we develop ourselves, how we relate to others, how we relate to society and to the world around us, this is all realized through culture which is changing and dynamic, so although AI presents a particularly rapid and potentially very powerful transformation in our culture, change is always with us. For this reason, I'm wary of those who count the number of technological 'revolutions' we've had, claiming AI is the 'next big thing', perhaps the biggest of the lot; if you look at lists of the numbers of different technological and cultural revolutions humans have undergone, there's a slightly different list each time.

It may be hard to see the change as it's happening, and to predict what its consequences will be. AI may present us with an 'event horizon' (Vinge 1993) where we cannot see the change coming, or a 'phase change' which is not apparent to those immersed in it (Wallich 2008).

6.1.3   Social, Cultural and Technological Change Is Multifactorial

Consider this example of social change: which came first, effective contraception or the emancipation of women? Which was the greatest factor, the realisation that enlightenment ideals of reason also apply to women; female suffrage; the work of women in male occupations during the First World War; control over fertility; or something else? Books can and have been written on this topic. And consider, too, the example of antibiotics—a great boon to the treatment of disease. But without the more humble advances of sanitation, pest control, the mechanisation of agriculture and the transportation infrastructure that allowed massive populations to be clean and well fed, antibiotics may well have been a footnote to the history of the ravages of infectious disease on urban populations. The rise of AI, then, likewise built upon the gains of the humble yet noble sewage system, as well as a host of other factors.

And note, too, the complex changes to our ethical notions. Many are concerned that the development of AI takes heed of gender parity and avoids reproducing patterns of sexism (Fessler 2017). These aims are historically not a given, but themselves derive from complex technological and societal changes, and indeed, their precise instantiation is still the subject of often heated debate (Mundy 2017). Had AI been developed in a society which saw different roles and had different rules for men and for women, a top research priority of engineering funding councils might have been the creation of labs dedicated to ensuring that there are enough Robot Wives for all men to practice polygamy.

And why is it necessary to point these things out? Because if we forget that these ideas and values have been achieved through effort, we may take them for granted as a given, and forget to guard them, forget to watch, while they slip away.

Exactly how technology develops is affected by factors beyond technology. Technology does not alone *dictate* what the future holds in store for us, much less what we think and feel about what the future holds in store (Mackenzie and Wajcman 1999). As changes occur, so too do anticipated responses to ethical questions, and there is no reason to suppose that we will reach any consistent or predictable set of ethical judgements. This may be especially so for AI, insofar as AI impacts upon how we think, how we relate to each other, and on vital elements of society such as modes of production. Hence, positioned from where we are now, it may be very challenging indeed to anticipate adequate ethical responses.

The code of ethics of the Association of Internet Research (AoIR) explicitly looks at how information technology is changing societal relations (AoIR 2012). This is potentially very useful for AI to develop further understandings of the complex interactions between AI and people, and at how the use of AI might change our very understandings of ourselves and our communities upon which our value judgements are premised.

To what extent can a professional code of ethics handle such complex questions? We are facing a perennial problem, that of trying to find order, meaning and benefit from changes we don't quite understand. In regard to AI, questions are presented in forms which are perhaps particularly acute, particularly hard to spot and to fathom. At the risk of sounding like a scratched record (remember them?), we need to bring all the resources we have to hand to bear on this, and to involve as many people as possible.

6.1.4 Change and Moral Uncertainty

In a rapidly changing context, just how integrity should be cashed out in any particular context becomes very unclear (Bowden and Surma 2003; Cohen et al. 1992). If there are rather different sets of values between different social groups, as for instance seems to be the case with the value of privacy, then that sets a big problem for how we navigate these in a fair and equitable manner (Nissenbaum 2010).

Political and individual differences: Note, too, that in the example, much of the debates about privacy are highly politicised and depend deeply on, for example, how far one considers there to be a threat from cyberterrorism and from terrorist groups in general, and how far one trusts current governments. So debates will depend greatly on individual differences in relation to the detection and assessment of threat, which in turn will depend upon individual differences in personality and in personal history. Where issues are understood to engage with political issues, we need to consider the place of professional bodies. This is a complex question, but a professional body often purports to be, or strives to appear to be, relatively neutral with respect to political affiliation and issues, unless these are of direct concern to the membership in their professional role. Certain politically charged questions might then be considered to be inappropriate, or beyond the scope of a professional body to comment on. Yet these might be among the most important questions facing that profession. Questions of philosophy, economics, society and culture all are raised directly in attempts to regulate the development of AI.

There are certain dangers lurking. One is that political or other value concerns will be present in the pronouncements and ethical codes of a professional body, reflecting an assumed orthodoxy that may not represent all the members, and that may not have received adequate consideration and debate. Where professionals come into their own is in understanding their area of expertise, rather than bland general statements. For example, among the vexed questions facing technological advance in general and AI in particular, is whether, and if so how, new technology may deepen existing social and economic divides. Research shows that the introduction of medical testing tends to increase health disparities, since such technology will be disproportionately utilised by the already healthier educated middle classes, but also that the picture is very complex, for simplifying technologies can increase uptake by less privileged groups (Goldman and Blanchard 2015). Careful research and detailed policy applications are needed, especially if we are to avoid bland hand-waving gestures.

What Is a Parent? How Technology Is Changing Our Concepts. And Does It Matter?

Developments in science and technology frequently present us with questions of boundaries and definition for central value concepts. For example, the need for increasingly precise definitions of death have since the twentieth century raised profound questions which we did not need to face until certain things became possible, such as life support technologies and organ transplantation.

Here's a question: Is 'parent' a descriptive term only? Or is it a value category that helps to shape the patterns of our moral thinking? Values terms are not simply broad and obvious ones like 'good', 'right', 'pain'; 'happiness', 'health', 'horror'. One task of ethical thinking is to consider how technology might impact upon concepts which are not overtly 'ethical' but which are imbued with value and significance.

There have always been divergences between biological parenthood and social parenthood. The question of who a child's 'true' mother is in indeed as old as King Solomon himself. (1 Kings 3, 16–28). It was with the advent of reproductive technologies, coupled with changing societal attitudes to same-sex couples, that questions about the precise definition of what it is to be a parent arose in their present form. Indeed, the very presence of regulatory bodies which formally specify rules around parenthood is itself one driver for changing definitions.

In the UK, there are complex procedures for determining the parents of a child born from reproductive technologies. Interestingly, this includes techniques which are not especially 'hi tech', such as sperm donation, as well as technically complex procedures. The definition of who counts as a child's parents makes no essential reference to gametes. The default position is that the mother is the person who gives birth, but the final answer will depend upon a number of factors which include social and legal relations; the legal mother may have neither given birth, nor provided the egg. There is a very complex decision chart on the Human Fertilisation and Embryology Authority (HFEA) website, far too long to reproduce here (Legal Parenthood 2008).

A few points. *First*, this is a clear example of how technology, along with changes in social attitudes, has acted as an impetus to uncouple a central social concept from the biological or 'natural' world. Some advances in AI, at least those concerning more direct forms of human enhancement, may promise the same.

Second, regulatory bodies such as the HFEA often call for public debate and consultation on such matters. However, this particular change does not seem to have occasioned very widespread public discussion, perhaps puzzling since changes seen as alterations of 'nature' are generally met with at least some vocal opposition.

(continued)

Third, this shows how technology may be affecting central social and value concepts, even for those who are not users of the technology.

Fourth, there's no necessity for both a mother and a father; there can be simply 'parent one' and 'parent two'. So, since we are departing from nature in other ways, why follow the pattern of nature in providing a child with two parents? It's always been the case that sometimes, the identity of the father is a mystery. But now the way is, in theory, open to further possible changes, even three or more parents. Why not?

The lesson for AI? This commentary makes no judgement about the nature of these changes; but we do need to understand what's happening. Watch out for similar such incremental changes in terms that express social values in relation to AI.

6.2 Social, Cultural, Economic and Technological Change: The Example of AI and Employment

It is widely accepted that advances in AI are already having profound effects on employment, and that this will continue to be the case in the future. Questions concern unemployment caused by AI, the distribution of work and therefore of wealth, as well as the changing nature of the employment market, and changes to education and training (Mokyr 2014; Brynjolfsson and McAfee 2014; Brynjolfsson et al. 2014; Frey and Osborne 2013; Glaeser 2014; Shanahan 2015; Nilsson 1984; Manyika et al. 2013). So, how should codes of ethics in AI address this, given that this raises questions far beyond the remit of AI researchers per se?

Complex political questions are involved. Many recognise the complexity of the employment issues AI will bring, but sometimes commentary calls somewhat simplistically for solutions such as a Universal Basic Income for the unemployed. Yet, a brief glance at bloody twentieth century history tells us that calls to distribute income 'fairly', taken to certain extremes, have not worked out terribly well so far around the world. Karl Marx himself of course has already envisaged that robots will eliminate work in a communist utopia (Marx 1858). One take-home lesson however is how much governments will have to be involved in addressing the fall out of AI, as the recent reports of the US, the UK and EU make clear (European Civil Law Rules in Robotics 2016; Robotics and Artificial Intelligence 2016; Preparing for the Future of Artificial Intelligence 2016).

As so often, these debates are not new. Debates inspired by the industrial revolution about the nature and value of work (Morris 1893; Ruskin 1904) preview current debates about the replacement of human labour by AI, and both old and new debates raise questions such as the link between employment and unemployment and happiness, and the nature of leisure (Hetschko et al. 2014; Russell et al. 2015).

Looking to the past can be useful; having a historical perspective is akin to getting an interested outsider to comment on our current debates.

What is work? In considering how AI is affecting the workforce, we need to consider fully the meaning of work. It may seem obvious to replace onerous, boring, time consuming or dangerous work with a machine, eliminating what William Morris called 'useless toil'. But, in complete contrast, is the view that work has value, and that working gives the worker purpose and meaning. AI enters the debate, threatening, or promising, to replace work which is more complex, which to many, would appear more archetypally human and meaningful. However, we need a note of caution here, for the question of the value of work and its place in individual and societal life is complex. Many measures of meaningful activity make certain intellectualised assumptions which require critical scrutiny. (Boddington and Podpadec 1992).

There are multiple, complex benefits from work: Social connection with others; a sense of belonging to something larger; a routine; a reason to leave the house; an identity; an organisational structure wherein one can attempt to advance one's position; recognition in the form of a salary and perhaps bonuses or 'staff member of the month' awards; and health benefits. There is clear empirical evidence of the health benefits of work, especially if one is given responsibility and agency (Marmot et al. 1997; Wilkinson and Marmot 2003).

Measuring the impact of the changes that AI brings to employment will be extremely complex. Consider this example. One imperative of AI is that of speed and efficiency. But the question of whether or not technologies, including AI, are even 'saving' us time involves complex questions about how we conceptualise and measure time and speed (Wajcman 2008). This further underlines the need for those with specific expertise to assist AI professionals in developing AI which benefits human flourishing. Note, that a question such as this about the nature of time and speed might well not even occur to you, had it not been pointed out by a subject matter expert.

And consider what profound questions are raised. Some envisage that AI will even take over creative work (Bostrom 2014). So what for humans then? The question that the development of AI thus pushes us towards is—*hang on, why are we here*? This is in many ways the polar opposite to the questions raised by our possible extinction at the hands of a menacing aggressive AI. If you are fighting for your life, the question of why you are alive may not occur to you. But if your life stretches ahead filled with endless leisure, if the world's problems have been fixed, and if a machine can write a love letter to your beloved better than you can, and a robot can fulfil your beloved's sexual desires better than you can, if the version of *War and Peace* written and acted by robots is better than the BBC's adaptation, you might start to wonder why you are alive. No mere code of ethics for AI can address this question adequately. But such a code needs to be aware that the question is there for the asking, and that others will, and are, asking it. We are basically raising that thorny philosophical question—what is the meaning of life?

Artificial Unemployment: A Sci Fi Tale About the Future of Work
Here's a sketch for a sci fi tale about a possibly not very distant future. It's a simple tale, and other stories could be told. But it is extrapolated from some known features of human and group psychology and behaviour, so it's not entirely far-fetched. The point is to give a fast and dirty peek at how many factors are involved in the unfolding of our future, and hence how troublesome it will be to devise codes of ethics and regulations to anticipate and ameliorate such changes.

Those laid off by AI are going to be those in professions such as accounting, legal research, and teaching. (Have you ever noticed how writings about AI and unemployment often seem to imply, '*In the past it was low grade jobs that were lost, oh dearie me, now it's our turn*'?) The jobs that remain will require high levels of training and skills, as well as those elements of caring work where robots aren't preferred, and artisan work—which only the rich will be able to afford. This means that the significant portion of the population with low capabilities may simply be permanently unemployed, in addition to many higher skilled workers who are no longer needed.

One group likely to struggle very badly without the structure and meaning provided by work are those who are high on the personality trait of orderliness. Research shows that such people are likely to be politically and socially conservative, hence constitutionally reluctant to embrace change, especially chaotic change. When employment goes up, an obvious response is to identify and target groups who are 'taking someone else's jobs'. So let's say then that in this sci fi scenario, since it's the conservative types who are most distressed by unemployment, there is a widespread call for women to leave the workforce and go back home. On some futuristic scenarios, the women will be okay, since they'll have the jobs in caring industries. But on this scenario, they've cracked carebots, and that, in conjunction with the force from some quite frankly livid alpha males, large numbers of women are forced, nagged or discriminated back into unemployment, a.k.a. 'the home'.

But then displacing women back into the home might also lead to a rise in the birthrate—because why be a stay at home mum your whole life and only have two kids? There's just not enough to do, you may as well have four or five, especially as their income is guaranteed under the new Universal Basic Income, although given the employment situation, you're likely to have your kids hanging around in your basement forever. Meanwhile, since there will still be so many without the social stimulation, the validation and the sense of purpose that even a suboptimal workplace can bring, pharmaceutical companies will be making a mint from the production of SSRIs and anti-anxiety meds, prescribed to you by your robot physician, whose finely-tuned empathy seems like the ghost of a smirk when you reflect on the fact that Robo-Doc will never, ever need such meds itself. Watch out for the new, lucrative

(continued)

disease of RDD, coming to DSM VII—robot dysphoria disorder. It will make pharma companies billions.

Those on UBI are likely to be unimaginably worse off than the tech guys at the top, with no visible rungs to the top. We know from research that increases in social inequality lead to a rise in crime and aggression, principally among men who vastly outnumber women in criminality, especially violent crime. After all, there were already reports in 2014 that people were stoning buses carrying Google employees to work (Rotman 2014). These won't on the whole be the same as the orderly types; they might even be people who initially embraced unemployment. Those confined to a lifetime on UBI will have no way of increasing their social status, a major human motivator, or advancing themselves through employment. However, other tried and tested ways of advancing social status on a low income include gang membership and criminality. This will include cyber-crime of course. Most will not do this; but it only takes a few to make life hard, and a small increase to make life hell. Of course, this is likely to manifest patchily, and the elites won't be living anywhere near the trouble.

Meanwhile, the men who do have jobs will have their pick of females. Face it, as one tendency in mate choice among others, it's true. 'Don't you know that a rich man is like a pretty girl? You don't marry her just because she's pretty. But, my goodness, doesn't it help?' as Lorelei Lee artlessly told us in *Gentlemen Prefer Blondes*. This will worsen the aggression of many of the have-not males. Other women may fall prey to sexually aggressive gang members. (It's happening now, so why would it stop?)

And everyone on UBI will be utterly dependent on a massive state. Apart from those at the top, and those living in the shadows in criminal gangs. This will only increase the attraction of gang life, as links between criminals and organised political protesters of neo-Luddites are formed. Sensing unrest, the state starts to crack down on 'citizen journalism' and the channels through which it is spread, aided by the Top Feeders of AI super-geeks and some very powerful algorithms. Tim Berners-Lee had announced in 2017 that he was working to address these issues, but was proven powerless in the face of the world wide web that he himself had started. The populace is utterly dependent on Google and Facebook for its news of what the state is up to.

Some utopia.

6.3   Regulating for Whom? The Global Reach of AI, Universalism, and Relativism

Universalism in ethics, relativism and universal human rights—a baseline for progress? Some issues in AI are going to have a further reach than simply one particular region of the world. At the very least, there are concerns already expressed about harmonisation. The EU Commission on Legal Affairs has broached this openly in concerns that if the EU does not regulate robotics, other parts of the world will introduce regulations and standards which the EU will be forced to adopt or align to (European Civil Law Rules in Robotics 2016). At an individual level, perhaps we just need to make sure that everyone can use AI in ways that accord with their own personal values. But to think of ethics as no more than the provision of 'free choice' is to ignore the many ways in which what we do affects others. How do we move towards trying to find the 'right' answers in codes of ethics, whilst acknowledging cultural difference, differences in ethics and mores in different parts of the globe? Even calls for minimum standards such as AI which complies with human rights must answer the question—with which charter of human rights?

A notion of universal human rights is often used as a base point of common agreement in discussing ethics. This can and is used as a touchstone in addressing both change, and concerns of diversity in respecting the rights and freedoms of others. For example, in its discussion about how to protect humans from harms caused by robots, the EU refers to the Charter of Fundamental Rights of the European Union of 7 December 2000, which set up human dignity as the foundation of all rights, with Article 1 stating that 'dignity is inviolable' (Directorate-General for Internal Policies 2016, p 22). Codes of ethics in AI may consider using human rights as a foundation stone of value. Rights have a valuable place in addressing ethical issues, although alone they are insufficient for a full account of ethical value. However, this aside, the appeal to universal human rights is not without its problems. Here are some brief notes on some of the major questions.

Firstly, human rights need some justificatory basis. What makes them universal? One answer would be their universal recognition. But, human rights are not necessarily universally recognised nor always undisputed. It's often overlooked that a claim to universalism in values is not recognised universally.

Human rights may be argued to be universal on the basis of a claim that all humans are of equal moral worth and dignity. But, (those who pin their faith on science alone, beware), there is no simple or direct *empirical* evidence for this claim. And there is dispute about this. Some consider that criminality, for example removes a person from the universe of moral equals. The Cairo Declaration of Human Rights posits that human dignity can be enhanced, implying that we do not all have equal dignity: 'The true religion is the guarantee for enhancing such dignity along the path to human integrity' and that 'no one has superiority over another except on the basis of piety and good deeds', something that many others would strenuously deny. (Cairo Declaration on Human Rights in Islam 1990). The very notion of dignity is usually used precisely to flag an unmovable moral equity between people, on the basis of which the value of a human being cannot be traded off against any other value; this notion of dignity simply does not come in 'luxury' and 'economy value' versions. Note that the Cairo Declaration was originally signed by 54 countries, so cannot simply be dismissed as a mere anomaly.

Human rights may be argued to be universal on the basis of a foundation in universally shared human needs; but it is surprisingly hard to articulate what, precisely, these are, in a way that does not smuggle in assumptions and cultural bias (Reader 2005).

Human rights may be said to be fundamental in the sense that other rights and indeed any meaningful life at all depends on these rights. The right to life is the prime example of this. But questions remain. As well as a negative right not to be killed, does this entail a positive right to the means to sustain life? The questions about life extension and enhancement that some aspects of AI may raise also raise complex questions here.

Other purported rights may claim universality on the basis of what is minimally needed for successful agency in the world, or for a well-functioning open society. Free speech may be one such alleged universal right. But there are many questions about the limits of free speech, and different societies around the globe have very different attitudes to it, and to its value.

Why is the basis for human rights an issue for AI? The potential global reach of AI heightens these issues, especially where it is entwined with the use of communications, which are indeed a major factor in debating and achieving human rights. As I write, there is a heated debate going on about whether Facebook is or is not planning to assist the Pakistani government with removing 'blasphemous' content on Facebook (Hassan 2017); AI could potentially be involved in this or similar activities. Much of any such 'blasphemous' content is concerned precisely with arguments for human rights which are guaranteed elsewhere in the world. Whatever the truth of this particular matter, it serves to illustrate a general and important issue that may affect applications of AI: it is hard to overstate the importance of such a matter; blasphemy may attract a death sentence in Pakistan (Siddique and Hayat 2008).

Free Speech Debate: An Example of Working Through Cross Cultural Debate

There is no escaping the fact that many applications of AI will have international and indeed global reach. So how can we have meaningful debates about the ethics of AI across cultures?

I'm afraid to say I am not going to offer a solution. But given that one of the crucial things we need to do is to start and sustain meaningful dialogue, an interesting example of a project that is actively attempting to find solutions and construct dialogue globally is the Free Speech Debate project (http://freespeechdebate.com/en/).

This is of interest to us because free speech is an important value; it's a value because of its vital contribution to the advancement of thought through open discussion and exchange of ideas, which I presume then is of interest to those in any scientific or academic enterprise, and perhaps especially to those who are pursing AI for the sake of maximising the intellectual and reasoning capacity available to humanity.

(continued)

Free Speech Debate is concerned with how technology, which may involve forms of AI in features like search engine optimisation and the filtering of results and communications, are affecting free speech. It also involves efforts to investigate issues of free speech and expression around the world. It is set up to invite involvement from people. It currently is available in 13 languages. (Of course, this is a fraction of global languages, but it's a start.)

Interestingly it directly addresses the question of universalism and whether this project amounts to 'Western Imperialism'. This is of course a highly complex issue, but the project does, for example, point to long standing traditions of free expression, debate, and listening to the viewpoints of others from around the world. "To block people's mouths is worse than blocking a river," says Duke Zhao to King Li in the fourth century BCE Chinese Discourse of States (http://freespeechdebate.com/en/the-project/).

Ten principles are presented for discussion: Lifeblood (on the importance of free expression), Violence, Knowledge, Journalism, Diversity, Religion, Privacy, Secrecy, Icebergs, and Courage. Many of these suggested principles balance different concerns and values. In this respect, they more closely resemble the loci of debating points than principles. For example, 'Secrecy' states: 'We must be empowered to challenge all limits to freedom of information justified on such grounds of national security'. The principle of reasoned challenge is one thing; denying the need for some government secrecy is something else, and the precise line will vary. Such secrecy can protect the lives of innocents as well as further the nefarious ends of rogue governments.

6.4 Diversity in Participation as Part of the Solution

Is drawing up codes of ethics a job for the morally virtuous, and if so, how do we identify these? The idea that we can identify the virtuous person may mean we end up taking on people of a certain type. There's a danger that people judged 'virtuous' might mean 'people like us' or 'agreeable people' and certainly, neither of these are equivalent to the possession of virtue. Moreover, people consciously interested in ethics are not necessarily more virtuous (Schwitzgebel 2009).

Rather, I suggest it would be a wiser strategy to attend to the diversity of those participating in developing codes of ethics. We have noted that any code of ethics in AI needs to deal with layers of uncertainty: uncertainty and disagreement on ethical values, uncertainty about technological development, and uncertainty about how individuals and society as a whole might respond to technological developments. The lack of a clear and undisputed methodology for advancing ethical debate was noted in Chap. 1. In its absence, including a broad range of views and approaches in debates about ethics and in drawing up and implementing codes of ethics is an important desideratum. However, at the same time, note that diversity 'quotas' of different groups is not what I have in mind: for one thing, humans come in so many overlapping categories that innumerably many such groups could be delineated. It's diversity of opinion, of experience, and of thinking style, in groups as a whole, that's important.

We have also noted how, especially in the context of AI, there may be silos of thought and opinion, with people of similar backgrounds and education taking a very prominent lead. We've also noted the concentration of power and resources into a few prominent hands in the field of tech and IT in general and in AI in particular. Those working in AI tend to have certain demographic profiles (Bass 2017). Margaret Mitchell, a researcher at Microsoft, refers to what she calls the 'sea of dudes' problem (Clark 2016). Research shows that the wording of job ads in tech may discourage female applicants, with most pronounced tendency in jobs in machine intelligence (Snyder 2016). It makes sense to hold as an ideal for diverse participation in the development of codes of ethics in AI.

But what do we mean by 'diversity', and why, precisely, does it matter? There's known social influence bias in how we think. Technology may make this worse, which again gives us reason to consider this particularly in relation to IT in general and AI in particular; the use of online resources and social media may herd us into like-minded cabals of opinion (Muchnik et al. 2013; Pariser 2011). There is also considerable work on social network theory which likewise demonstrates such significant effects (Christakis and Fowler 2010; Jackson 2010). It would be arrogant to assume that academics and the other opinion leaders working on the ethics of AI are immune to this. Indeed, there is considerable evidence that academics tend to have atypical attitudes to ethical and political issues (Haidt 2013). One arena which is attempting to encourage diversity of viewpoints in debate is the Heterodox Academy (Heterodox Academy 2016).

There are reasons of fairness for including the voices of diverse groups in discussions of ethics. A considerable body of work in public engagement with policy making and with research and development deals with these issues of wider engagement and inclusion (O'Doherty and Einseidel 2013). This rationale for inclusion focuses on the justice of getting voices heard, and in forming policy and developing research and innovation which caters effectively to people's needs. Work on 'standpoint epistemology' argues that those with certain experiences or identities may have privileged, sometimes exclusive, access to certain insights and understandings relevant to ethical inquiry (Campbell 2015). However, note that simply including a person from a certain group does not in itself guarantee that they are typical of that group, let alone that they in any way 'represent' that group.

Wishing for a good outcome gives reasons to consider diversity in group thinking. Research findings suggest that the problem solving skills of a diverse group outperform those of a group comprised of the most able individuals (Hong and Page 2004). Recent work shows that collective intelligence in group decision making is a factor independent of individual intelligence, and is correlated with the average social sensitivity of group members, the equality in distribution of conversational turn-taking, and the proportion of females in the group (Woolley et al. 2010). This gives reason to pay attention to the gender balance of groups, although the research so far is suggestive of the importance of social skills, rather necessarily gender *per se*. It's a reason to be concerned about the mix of minds, rather than focusing on getting 'the best' thinkers. Indeed, work indicates that greater cognitive sophistication often renders thinkers less flexible owing to their ability to dismiss contrary views (West et al. 2012), giving particular reasons for concern about ethics discussions formed entirely of like-minded subject matter experts and 'leaders' in their field (Hatemi and McDermott 2016).

This work in collective intelligence also finds some confirmation in recent work examining the notion of *metacognition* (Frith 2012). This is a process by which we monitor our own thought processes, taking into account the knowledge and intentions of others. This enhances and enables joint action by allowing us to reflect on and justify our thoughts to others. Individuals have limited ability to do this solo, but working in groups can enhance this capacity. Whilst metacognition would have widespread application, it is in ethics and in the justification of actions and decisions to others that it is of particular relevance. Here, psychological findings fit extremely well with long traditions in ethics which emphasise the importance of understanding our own motivations and thoughts, yet noting the difficulty of doing this alone (Butler 1827). Note that even such a foremost proponent of individual moral autonomy as Immanuel Kant recognised that autonomy requires that moral actions must be based on the right motivation, but also how difficult it was for us to know our own real motivations (Kant 1972). Work on self-deception likewise indicates the need for awareness of the distortions in one's own thinking in considering ethical issues (Tenbrunsel and Messick 2004). This means that any groups working towards developing and implementing codes of ethics must attend to the quality of their discussion and constructive, critical feedback.

Inclusion, group performance, and hierarchy: Humans are not just social animals, they are also hierarchical animals, but determining which forms of hierarchy and which methods of collaboration produce the best results in particular situations is complex. Work which shows the effectiveness of groups with mixed social dominance also shows the complexity of these effects, and suggests that more work is needed (Ronay et al. 2012). Other work shows that testosterone disrupts collaboration, with a tendency to overweight one's own judgements compared to that of others (Wright et al. 2012). Since testosterone levels vary not just between genders but within genders, again, simple recipes for good group construction cannot be necessarily drawn from these hints.

Likewise, other work demonstrates that in collaborating, it is important not just that you collaborate, but with whom within a group you collaborate (Smaldino 2014). Certainly we can conclude that paying attention to the nature and quality of in-group communications will be useful in our attempts to produce the best outcomes in developing codes of ethics and in the thinking of ethics boards about particular cases. Trials of different models may be useful, and further work is needed.

What about gender? The current literature shows mixed findings for the value of the presence of females for team performance, with differences attributable to context (Bear and Woolley 2011). It's suggested that where there is already a reasonable gender balance, the presence of women enhances team performance, but where there is an imbalance with a preponderance of men, then sub-groups with a fairer gender balance *suffer from the negative stereotypes of women in that field* and do not perform so well. This could be problematic in some technical areas, where gender parity is either far off, or may never be naturally achieved, since it is entirely possible that in some areas of human activity, even absent any barriers to participation, one gender may have a greater interest in participating than the other (Schmitt et al. 2016).

At the moment, relevantly to our considerations, AI, computer science, and also professional philosophy, have more males participating than females (BeeBee and Saul 2011; Aspray 2016). This places us in a possible conundrum. The presence of women in a group may enhance group collaboration, but not necessarily enhance group performance, *where negative stereotypes of women persist*. Hence, where ethics exists as an activity within a male dominated area where there are negative stereotypes of women, inclusion of women within the ethics endeavour might, it could be speculated, act to produce negative stereotypes of the enterprise of ethics, especially if the presence of women merely enhances group collaboration, but not group performance, owing to persisting negative stereotypes in some fields.

Leadership: Recent work shows that narcissistic individuals are *perceived* as effective leaders, yet in reality, narcissistic individuals inhibit information between group members and inhibit group performance (Nevicka et al. 2011). The inhibition of sharing information would be especially relevant in the case of ethics, and the importance in ethics of effective self-reflection likewise suggests that the trait of narcissism in group leaders would be especially unwelcome. It seems that Plato had a good hunch when he argued in the *Republic* that only those who do not wish to lead should be allowed to do so (Plato 1974).

Findings from personality theory and moral thinking: Recent work in social and moral psychology, and also in personality theory, indicates that individuals with different personalities place greater emphasis on certain core moral values, and may describe ethical issues in different terms. This work has become well known in investigating the foundations of liberal and conservative thinking. This is particularly important given the finding that academics tend to have particular personality profiles and to lean left of centre. This is also true of entrepreneurs, so those in AI tech start-ups may share similar leanings. It would therefore be valuable to ensure that in AI had a wider range of participation. Note also that liberals tend to emphasise a narrower range of values than conservatives; this may limit ₁the range of ethical debate and discussion (Graham et al. 2009).

There is a deeper reason for doing this other than simply evening out the political spread of opinion. Personality research shows that some individuals score high in Openness, seeking novelty and new experiences; these tend to be left leaning. Others score higher in Conscientiousness, and in the subclass of this, Orderliness. Such individuals have a greater respect for tradition and for authority. Left leaning individuals tend to be much more tolerant of disorder (Graham et al. 2009).

The task that faces us in the rapid introduction of new technology is precisely how to balance the unknown, the novelty of AI, and the disorder to individual lives and society that this might bring, with traditional values that might be lost (Peterson and Flanders 2002). Having a group that only valued tradition would bring very different answers to a group that cared not a jot for traditional values and loved the exciting allure of the new. A balance between these two, with attempts to ensure fruitful dialogue and understanding, would surely be the best option, given what's at stake.

Diversity and hierarchy in tech: This discussion so far has been premised as far as possible on research findings to draw conclusions about the make-up of committees or groups considering the ethics of AI and in developing codes of ethics in AI. I'd add also that the research on social epistemology and the operation of hierarchies suggests that it would be useful to include people from all levels and all roles in an organisation (Goldman and Blanchard 2015).

And I have one last suggestion. I argued in Chap. 3 that in ethics we need close attention to all relevant issues, yet that the sparkly allure of technology can help to distract our focus from other issues. This gives good reason for consultation and close discussion with a range of affected people. It also, perhaps paradoxically, gives some reason to include in developing codes of ethics for AI, people who are pretty indifferent to technology, since they may be less likely to be distracted by it.

Chapter 7
Some Characteristic Pitfalls in Considering the Ethics of AI, and What to Do About Them

Abstract Those developing codes of ethics for AI must of necessity consider the ethical issues that AI presents. There are some common pitfalls and gaps in argument to watch out for here. A full treatment of this topic would take much longer, but this chapter simply aims to alert readers to some of the main traps to avoid. There is always a balance between abstract and concrete thinking in ethics. Work in AI and ethics may concentrate too much on the idea that what distinguishes humans is their intelligence, and subsequently, idealisation or oversimplification of what is involved in both human and machine agency may occur. There may be different expectations for human and machine agency which are present but not fully articulated. This can have concrete and deleterious impacts upon any ethical conclusions which are drawn. AI is used to enhance or replace human agency. This means we must pay attention to questions about the boundaries of human agency and 'normal' human functioning. There needs to be careful consideration of different cases, given the varying nature of AI. The impacts of AI may not be just on its immediate use, but further afield within complex social systems, and careful attention should be paid to this. Lastly, clarity of language and of definitions is frequently an issue in AI; common language may mask deep disagreement.

Here, I consider some characteristic pitfalls that may be found in attempts to address the ethics of AI. Again, this is not intended as a comprehensive list of problems, nor to suggest that such difficulties are inevitable; neither is it intended as a failsafe manual of how to avoid problems. The broad questions I examine all relate, in some way, to the ways in which AI characteristically enhances or replaces human judgement and agency.

7.1 The Idealisation of Human and of Machine Agency

7.1.1 The Abstract and the Concrete in Ethics

There is a tension in ethical thinking between abstract, general principles and goals, and the concrete particulars of cases. However, in considering general ideas in

© Springer International Publishing AG 2017
P. Boddington, *Towards a Code of Ethics for Artificial Intelligence*,
Artificial Intelligence: Foundations, Theory, and Algorithms,
DOI 10.1007/978-3-319-60648-4_7

ethics, we can allow our thought to become too abstract and hence we can miss important detail necessary for appropriate application. There are many ways in which this broad problem has been tackled by moral philosophers: for instance, John Rawls' famous notion of Reflective Equilibrium presents a methodology for addressing the balance between theory and concrete particulars (Rawls 2009).

It is by no means inevitable that overly abstract or idealised notions of agency will find their way into discussions of AI and ethics. Indeed, this is a great potential strength of work in AI: that decision making and the behaviour of machines has to be thought about in considerable, concrete and applicable detail. This then can help guard against over-abstraction. Concrete, practical work in AI, for example in the area of robotic-human interaction, can itself uncover various ways in which assumptions and idealisations about humans and human agency can cause problems. In a recent interview, Anca Dragan, who runs the InterAct lab focusing on algorithms for human-robot interaction, remarked, 'We have to stop making implicit assumptions about people and end-users of AI, and rigorously tackle head-on, putting people into the equation' (Conn 2017b). Nonetheless, the focus in AI on agency and intelligence can at times nudge us into an overly idealised or abstract approach to ethical questions.

7.1.2 Artificial Intelligence, and Intelligence as the Hallmark of Humanity

Indeed, hype around AI can veer towards idealisation and simplification. For example, focus on artificial *intelligence* might lead us to overemphasise intelligence as humanity's main feature. It's common for those who avidly advocate AI to imply that there is some upward trajectory of advancing intelligence, an arc of moral progress, and that AI—artificial *intelligence*—is the next step to the progress of humanity—or of transhumans or posthumans. 'I regard the freeing of the human mind from its severe physical limitations of scope and duration as the necessary next step in evolution', states Ray Kurzweil (Kurzweil 2001).

But note, such thoughts often rest implicitly on teleological accounts of human evolution. Scientists are usually better known for considering evolution a product of blind chance, whereby species which don't adapt to changing environments simply die out. 'The necessary next step in evolution' implies that there's been some progress in evolution, but not enough; it's as if Nature herself, who fashioned us from inert matter, is now prompting us to wrestle evolution from her own amateurish hands.

Note, too, that such accounts tend to focus exclusively on intelligence as the factor behind humanity's current state of progress. Yet, theories of human evolution point to many other factors; sexual selection, which is a large factor in human evolution; critically, our social nature, including pair bonding, and the operation of dominance hierarchies; and quite possibly, religion (Barrett et al. 2002).

Including these factors in our considerations alongside intelligence may enrich our understanding of what constitutes progress for humanity. Two points for now. Firstly, recall our discussion of the question of work. One major question raised was how we would deal with the issue of the meaning of our lives with large scale AI-driven redundancy. A view of human life, and human progress, based on an account of intelligence alone—even on a wide notion of what 'intelligence' is—is going to be limited. Recall, for example, the question of the value of different sorts of work. Humans have very wide range of abilities and values, and a large repertoire for squeezing meaning out of life, but we are unlikely to be indefinitely malleable.

Secondly, these other factors in human nature are utterly critical to any account of ethics. Aristotle was not just a philosopher but the world's first biologist. He understood well that not just our intellect and our reasons, not just our emotions, but also our sociability is key to understanding human nature, and hence a key to understanding the 'good life for man'. 'For in the case of human beings what seems to count as living together is this sharing of conversation and thought, not sharing the same pasture, as in the case of grazing animals' (Aristotle 1999) Book IX ch 9. Evolutionary biology is catching up with Aristotle; our social natures have been key to our evolution: look, our large brains could never have developed without the empathy and complex society needed for care of the human infant, born helpless only halfway through gestation; in turn, much of our brainpower is concerned with social skills (Morgan 2011). If we see AI as human progress, if we are concerned about the ethics of AI, we must guard against a simplified attention to bare intelligence and to idealised, isolated individual agency.

7.1.3   Idealisation and Overreach Often Applies in Thinking About the Ethics of AI

Since AI has potentially very wide reach and there is concern about having our lives influenced in every which way by intelligent machines, there is a tendency to consider that the ethics of AI has to cover 'everything', so that we have to 'solve' ethics first. Hence, for all but the most cheery optimist about doing this (maybe someone who never picked up an ethics text book, nor ever watched the news), prospects may seem gloomy. Yet, at worst, this would only be an issue for a form of AI that really did affect everyone, and really did affect all areas of life. For many or most AI applications, certainly at present, there will be limited reach, and hence, the ethical questions, including the question of community agreement, is to that extent contained. There may be no need at all to fix the bigger, global ethical questions first; or at least, we may make some useful progress without this.

7.1.4 Idealisation in Thought About Autonomous Vehicles

We may think about agency differently in the case of human beings, and in the case of machines. This may be done inadvertently. This can mean that we have different expectations of machines; and this can infect our thinking about ethical issues.

There is reason to think that autonomous cars of the future will be safer than human driven cars: because if they're not safer, they won't be accepted; we are likely to be less forgiving if a machine kills us, than if a human being does, for a variety of reasons.

There is a general problem with measures which increase public safety. Realistically, these can never be perfect. The people who are kept safe are statistics. The people who are killed or injured are visible. Who reading this knows for sure that they would have been run over and killed, were it not for advances in vehicle safety? As it's been stated: "If self-driving cars cut the roughly 40,000 annual US traffic fatalities in half, the car makers might get not 20,000 thank-you notes, but 20,000 lawsuits" (Russell et al. 2015).

We may also feel that a human driver, otherwise competent and alert, faced with a vehicle collision in which a bad decision was made under great duress, should be forgiven. We are much less likely to 'forgive' a machine that does this. This is partly, I suspect, because of how an autonomous vehicle programmes in advance what to do in some crash scenario. This seems 'cold blooded'; recall the discussion in Sect. 2.3.8 about how anguish and slowness can be used as an indicator of moral sincerity. Whatever answer is preferred, it's going to help add clarity and nuance to the debate by considering directly how we are idealising machine agency and what happens when we substitute a human decision maker for a machine.

Note, too, a paradox: One main point of AI is to make decisions extremely quickly, and to careful formulae. But in ethics, it's often these very features of decision-making which occasion suspicion. This may indicate that trouble may always be on the horizon wherever machines are stepping in for humans in serious or even tragic cases.

We may also idealise in thinking about ethical issues because of the methodology used. The focus on 'trolley problem' type approaches to the ethics of autonomous vehicles, for example, may divert attention away from the wider context of the activity in question. For instance, focus on the precise number of people killed while driving in some abstract simulation might not lend itself to asking the bigger question of why you got in the car in the first place, given that you might end up killing someone. We take cars and road deaths for granted. Especially in those parts of the world which are heavily dependent upon private vehicles, it's a common attitude that humans have a right, a need, to drive. It's less likely that anyone thinks in these terms for introducing the new technology of autonomous vehicles. Individually, we also don't tend to get into vehicles thinking we are a danger to other road users. Collectively, although we don't want to be run over by another driver, we want to drive ourselves, and if the standards for driving skills were too high, too many of us would be ruled out. So, we're likely to be softer on humans than on machines in this regard. A downside of this is that safety concerns about autonomous vehicles may delay their use, even after they have reached the stage of being safer than human drivers.

7.2 Building Ethics into AI and the Idealisation of Moral Agency

We've seen how codes of ethics for AI need to build in an extra layer of complexity, one concerning the behaviour of machines. There are various ways of addressing the control problem. Could building ethical behaviour and decision-making into AI be one answer, as a strategy along with developing codes of ethics? Perhaps such codes may even incorporate as a desideratum of work in AI, a recommendation to build ethical behaviour into machines.

But does this make sense? A very brief snapshot of this idea is included here, for many of the pitfalls reveal simplification or idealisation of the notion of a moral agent. There are unarguable reasons to incorporate into the development and use of AI all steps to ensure safety, and to try to ensure that machine behaviour is consistent with ethical values. But the question about building in ethical decision making and action into AI goes further than this, for it concerns judgement in novel, perhaps unpredictable situations, where decisions and actions would be taken without any immediate human oversight; it goes further than simple alignment of outcomes with our ethical values, if it implies that it's the machine itself which is acting morally.

There are, of course, many forerunners, such as failsafe systems built into trains to cope for catastrophes, such as driver collapse. But these work in systems with limited capabilities. For systems of AI where decisions and actions may be made which might have far reaching, and perhaps hard to detect effects, the idea of building ways to make the decisions of the machine 'ethical' might seem a tempting possibility.

Eliminating catastrophe: Discussions around hard moral dilemmas are not just a hallmark of the ethics literature in general, but the ethics literature in AI in particular. So, in the absence of a complete specification of ethics, attempts to build ethics into machines may instead perhaps usefully be focused on at least trying to prevent appalling consequences.

But even here, it's hard to specify what these are. Is running over the baby a catastrophe, or is running over six 59 year-olds a catastrophe? Is it worse if the accident victim is left in a coma, or if they are killed? And, it turns out, *where AI is concerned many of the possible outcomes lie so far at the extreme limits of what we can imagine that they flip from 'wonderful' to 'catastrophe' like a Necker cube flips from one view to the other*. Is AI-induced mass unemployment the ultimate freeing of the human race—or is it a catastrophe? Is uploading my mind into a computer to gain eternal life (so long as you've bought a good policy for sorting out software bugs) a good thing? But this is what happens to victims of the Cybermen in Dr. Who, who are terrified at the prospect of being 'upgraded' to into a machine.

Attaining goals: It will be very hard to programme a machine to address ethical questions, unless we have a pretty clear idea of our value goals. But we lack such a clear view especially for such difficult to imagine, complex possibilities. So could we programme a machine to discover our 'true' goals? Well, on what basis would the machine work out our true goals? Well, perhaps either we or the machine can work out what our 'true' nature is. But ... do we even have a 'true nature'? And is our nature fixed then? And can this be something subject to empirical inquiry? This is an immensely complex philosophical question (Stevenson and Haberman 1998).

Moreover, even if we suppose we can create a machine that could determine our moral goals, this bootstraps up the problem in an unverifiable way. We would always need to be able to check that the outcome was ethical, by our own lights. Are we going to accept that, say, wife-beating was ethical after all, particularly if she's burnt the dinner *and* has sloppily applied make-up, just because we've got an app that told us it was okay? I hope not.

Outsourcing ethics: One of the central claims of this book is that ethics must always involve the possibility of development and of dialogue with others who have legitimate interests; perhaps they are affected, or perhaps they might have some insight to contribute. To outsource ethics to a machine that is not embedded in a web of such human dialogue is counter to all of this. And, should machines develop to a point of sophistication where they have as full moral agency as humans, although such a machine might have interesting things to say, handing over moral judgements to that machine is still outsourcing your ethics to another.

There is a serious problem with the whole idea of outsourcing our ethical judgements and actions to a machine, just as there is for outsourcing them to another person. In consequentialism, the only thing that matters ethically is the outcomes of our actions; this is an agent neutral morality where it does not matter how you reached a decision so long as it's the right one. So, you could, in principle, outsource your final judgement to an efficient machine. But note that this machine would be simply working out a decision procedure to implement a morality, and doing empirical calculations about how best to achieve a moral goal.

And on virtue ethics and on Kantian views of morality, you simply cannot outsource an ethical decision to others. You can't ask someone what to do and then do it, because to act as a moral agent intimately involves the quality of your motivation, and the nature of your judgement and decision making. You have to do the right thing, for the right reasons, in the right manner. Even many consequentialists are troubled by this, and try to work around it. And remember our discussion of the Nuremberg trials? The quintessentially bad excuse of the twentieth century was, 'I was only following orders'. That means that what is perhaps the most important moral insight of the twentieth century—upon which subsequent codes of professional ethics and laws have been built—is that we cannot outsource our moral judgements. It is a judgement of inalienable moral responsibility.

7.3   Replacing and Enhancing Human Agency, Boundaries and AI

One of the biggest questions facing AI is to consider the impacts of the enhancement or replacement of human agency by AI, and to start to analyse the multiple issues involved. There will be complex ethical questions; even if such developments are seen as beneficial, the question still remains of how such benefits are distributed. And, as we've seen, assessing the benefits of such complex and far reaching technologies associated with AI will be in any case, extremely hard. Moreover, simply trying to capture benefits and harms does not exhaust our moral discussions.

Hence, codes of ethics for AI need to be formulated in ways which permit and encourage the full complexity of questions about AI and human agency to be addressed. In particular, codes of ethics for AI research need to encourage research which actively investigates these issues where appropriate. Much research in AI already is looking at large complex systems, and hence could be a promising line of inquiry for including consideration of the ethical questions involved in displacing or supplementing human agency or human agents within such systems.

I noted earlier how thinking about ethical questions may be broader or narrower, and how those with different personality types may be more or less concerned with issues of boundaries in ethics. Given the central questions of how human agency, and even human bodily boundaries for some forms of proposed AI, affect boundaries, when we think about the ethics of AI, we should watch out for the tendency to reject or ridicule attention to boundary issues.

7.3.1   Case by Case Consideration Is Needed

Some AI may extend our capacities in incremental or relatively ethically insignificant ways. But the question of drawing the boundary between ethically significant and ethically insignificant will be contentious. We can see value questions about the enhancement of humans already operating in sport, and in medicine, with questions arising about the boundaries between curing disease or illness, and enhancing human capacities. As with AI, answers to such questions will depend upon ideas about what constitutes 'normal' human functioning, and the appropriateness of going 'beyond' this. There are problems about how to distinguish between incremental changes which have big effects—this is the question of the Sorites paradox of 'when does a few grains of sand become a heap'. One way of determining if a pile of sandgrains has reached the level of a heap, or if emergent properties are exhibited, is by looking further afield at the knock-on effects of the AI. A small change in AI capacity might have a substantial impact elsewhere in a system. For example, it might render a whole class of jobs redundant, and then lead to large institutional restructuring. But this will involve considering AI within its concrete, real world setting. Codes of ethics must therefore take note.

There is only time in a book of this length for brief indication of some of the issues. Let's consider a few examples. In 2016, a Robotic Retinal Dissection Device (R2D2) trial at Oxford was used for the first time to remove a membrane 100th of a mm thick from the retina of a patient. The membrane was distorting the shape of the retina. The robot was placed inside the eye through a hole less than 1 mm in diameter. The remotely operated robot eliminates tremors in the surgeon's hand. Such precision would be impossible for an unaided human hand. The device can perform movements as precise as 1000th of a mm (Parkin 2017).

It is hard not to see such a use of robotics as anything other than a great advance. The robot is controlled by the surgeon at all times, and extends human agency merely in terms of adding precision to human movements. Surgeons already have the ability to perform very delicate operations. And the purpose of the robot, to restore sight to as close as normal functioning as possible, can also be taken as uncontroversial—indeed, of great value.

However, consider a different possibility from a health care setting. There is much work on the potential for using robotics in nursing care, for example, for routine care work. Such work may be used to supplement human labour or to replace it. Working out its impact will be highly complex.

Take the use of robotics to assist with the feeding and toileting of patients on a hospital ward. Will patients benefit or not? For obvious reasons of privacy, a patient may well prefer robotic assistance with using the toilet to human assistance. But it remains to be seen if the same is true for other assistance. Routine care work provides opportunities for human interaction which may make a big difference to quality of life for patients, which in turn affects health outcomes, and may provide opportunities for exchange of useful information about the health status of patients. However, robots could also possibly record various details about patients, producing complex issues about data storage and communication within the hospital system.

7.3.2 What Kind of Questions Do We Need to Ask in Such Cases?

These are far more complex than simply assessing the benefit for patients. Hospital wards are intricate social environments where staff at different grades and functions operate in often varying local cultures, and where social hierarchies operate (Bridges et al. 2013). Within this social setting, there are complex lines of communication of morally relevant knowledge. In recent years, there has been an increasing professionalization of nursing, with nurses often using specialised equipment, and with routine bodily care more and more undertaken by lower status health care assistants (Twigg 2000). Technology seems to track social status. It's hard to predict what impacts there might be on relative status within a ward of the introduction of robotics for various aspects of nursing and routine care. And note that status within the ward is a critical element in how knowledge flows. Health care assistants may have particular, and useful, knowledge about patients, but they may or may not be shut off from ward meetings, and research finds that their low status, combined with the stigmatising nature of the bodily care work they perform, further isolates them as a relatively insular group within the ward (Lloyd et al. 2011). This has implications both for their own wellbeing, and for that of patients.

7.3.3 AI, Ethics, and Effects on Complex Systems

This is merely an indicator of a very complex issue. But in assessing the impact of AI entering a complex social system, it's going to be important to ask questions about how changes might occur elsewhere within that system, and to do so, one needs to understand how the system operates. I noted earlier the error of thinking of humans too much along the dimension of intelligence, a pitfall that might occur if we focus on bringing in artificial *intelligence*. We need to look at our social nature too. We need to look closely at bringing AI into human societies. One reason why I used a hospital ward for my thumb-nail sketch of the use of AI is because it highlights questions about social dominance hierarchies, something to which ethics needs to pay closer attention. Placing robotics into such systems may have unexpected effects, which may be trivial, or may be profound. We need to pay particular attention to how this might affect the transmission of information within a social system; crucially, this affects what issues are even seen as ethical issues, and how. Because those lower in social hierarchies, such as the health care assistants mentioned above, are less likely to be listened to, it may be especially useful to take such dominance hierarchies into account when considering an appraisal of the ethical impact of implementing AI within a social setting. Work in social epistemology could be useful here, and again, the kind of systematic thinking in which many experts in AI are adept may be useful. Codes of ethics might usefully consider explicitly addressing such matters (Goldman and Blanchard 2015).

Take note: Consider social systems. Consider social hierarchies. Consider the impact of technology on these and on the nature of communication. Consider how this might impact upon how ethical issues are uncovered. Consider whose views are least likely to be heard.

7.3.4 Pay Attention: Technology Can Hide, and Technology Can Blind Us

In ethics we need to consider not just what the right thing to do is. We need to consider how ethical questions are seen, how they do and do not come to our notice. The perennial issue in AI of how it supplements, enhances or replaces human agency means that we need to pay attention to what's going on with the human beings affected by the use of AI, and the complexity of human social systems may make it hard to see what impact the AI is having, without close attention.

There are additional questions about ethical visibility that arise with AI. AI takes many different forms. It may be so tightly and so invisibly embedded in complex technological systems that we don't even notice it's there (until it causes us some problem, perhaps); recall that once it's used, we may no longer think of it as AI. Contrariwise, AI may be whizzy, hi-tech, dazzling and exciting. Both these features—invisibility and prominence—are typical of AI, and both present ethical challenges.

We saw above in the brief discussion of robot camel jockeys how focus upon the robots as a solution to a moral problem might distract from considering other important aspects of the situation. Technology can over-complicate matters: Anti-Slavery International commented wryly of proposals to introduce robot jockeys into the UAE: '*This seems a complicated alternative to implementing fair labour conditions for adult jockeys.*' (Anti-Slavery International 2006). Technology which dazzles us can also prevent us from looking closer at other issues, as we saw above in discussions of how it seemed to be the lure of the technology which enticed owners to replace child jockeys, rather than any moral realisation of the wrongs of using children. Codes of ethics for AI need to consider carefully how these sometimes opposing aspects of technology—hidden, or revealed in full, chrome-gleaming lustre—may impact upon what ethical problems are visible, and what ethical solutions are sought.

7.4 Addressing the Increased Gradient of Vulnerability

We've seen how a distinctive issue for codes of ethics for AI is how the problem of control decreases the gradient of vulnerability between AI professional and others, which in turn threatens the authoritative base of professionals and of any codes of ethics. Attempts to address this are key to developing autonomous AI, and include discussions about how to retain meaningful control over AI, and indeed, what such meaningful control would even look like. It is obviously impossible in a book of this length to address what precisely to do about this. If I could answer the control question, this little book would be at the top of the Amazon best seller lists for sure.

However, the question of how to develop codes of ethics for AI, given the control problem, is somewhat different. It again reinforces the need for wide public communication and involvement.

It might be a crumb of comfort to AI professionals to see that there are similar issues elsewhere. In medicine the professional status of the doctor is being gradually transformed, eroding traditional notions of professional authority and causing numerous troubling questions for professional ethics.

Developments in technology, such as remote devices, some of which include AI, allow individual patients to collect and be in charge of a great deal of data concerning their own illnesses and health status, and, together with the rise of patient groups and more widely disseminated medical knowledge, this has led to the rise of the 'expert patient' (Department of Health 2001). This weakens the traditional expertise gradient between medical professional and patient, challenging the superiority, integrity and validity of medical knowledge, whilst notably validating one of the central values of medical ethics, patient autonomy. The boundaries and power base of expertise of the medical profession is also challenged now that much health data is in the hands of mobile phone companies rather than in the control of the medical profession, so new loci of struggle for control are arising as medical practices extend beyond the traditional clinical encounter; indeed, it is developments in AI, inter alia, which are producing such a challenge (Boddington 2016). However, there are not inconsiderable challenges to the authority of any resulting codes, since changes in the 'vulnerability gradient' both diminishes and modifies the power of the professionals and professional bodies who are producing these codes. It could be very useful to keep track of how medical ethics is dealing with such changes. This use of technology also raises questions about the potential loss of power of the medical profession. Changing patterns in knowledge and expertise between individuals and groups is a common feature of emerging technologies.

One response is to recognise the legitimacy of public concerns and to express these in forms such as widespread public consultations. These should be very welcome for AI.

7.5 Common Language, Miscommunication and the Search for Clarity

There is a pressing need for clarity of communication in any enterprise of investigating and developing ethics, and this is a particular issue with the ethics of AI both because of its technical complexity, and because of the need to add as much transparency as possible, given the difficulties with transparency in some forms of AI. All interested parties need to be able to understand the ethical issues, and so there's a need for technical language and concepts to be communicated clearly; but note of course, this need for communication goes both ways—those working in AI need to understand the concerns of those outside the field.

A particular problem in AI is that there are terms which are used in technical sense which are also in common parlance. Perhaps the prime example of this is the word 'autonomy'. This is used in particular ways by those working in AI; it's used in common speech; and it's used by philosophers. There may not be complete agreement between different uses of the word. And any misunderstandings thereby generated are likely to be important ethically; it's a concept to which great value is attached.

Despite the necessary calls for clarity on language and definitions regarding AI and ethics, a common vocabulary can mask disagreements. Only a depth of dialogue and an understanding of underlying background issues will reveal this. But don't be mistaken in thinking that all we need to do to improve understanding is simply need to come up with a robust and agreed definition of autonomy.

7.5.1 Common Language May Mask Disagreement: A Tale of Two Autonomies

Within ethics itself, autonomy can also be understood in radically opposing ways. You could literally fill an entire book case with material on this. Here I just illustrate briefly two contrasting approaches. This also serves to illustrate how deep debates about the ethics of AI are likely to go, and to warn how language may mask serious disagreements.

Autonomy may be used to signify human agency, responsibility and freedom, and it's frequently used in this context to flag the importance of allowing individuals to make decisions for themselves and to hold their own personal values, without any outside influence. It represents the rejection of external demands. It can be used to mean, 'I set the rules for me.' Recall that earlier, we discussed the view that morality only concerns how I treat others; that I can do what I want if it only affects me.

Yet we can trace back emphasis on autonomy in ethics to Kant's philosophy, in which it is an essential feature of human beings that they are rational agents, capable of autonomy. BUT note this: For Kant, to be a rational agent is to recognise the pull of rationality; we *participate* in rationality. Rationality gives rise to the demands of morality (Kant 1972). This means that in acting with full autonomy, we also act morally, motivated by our reverence for the Moral Law. And the Moral Law, based as it is on reason, gives universally applicable answers (at least in theory). This is freedom, this is autonomy, not because we are doing 'what the hell we like', but because we are acting in accordance with our natures as rational, autonomous beings.

To argue that autonomy means the rejection of 'external' demands and hence, that each person can do what he or she likes, is correct then, only if you ignore that for Kant and for his followers, the demands of rationality, and hence the demands of morality, are not 'external' to us.

Some tricky concepts that are likely to crop up: Note that there are many concepts where values are deeply implied, but which may not at first sight seem purely value terms themselves (Williams 1985). For example, consider the word 'parent' which we discussed earlier. We also looked at the notion of 'bias' which seems at first sight always wrong, but which on inspection, things are not so clear. The notion of trust is often used in relation to AI, especially in robotics, but again, trust is a two-edged sword. Adults grooming children for sex are very good at eliciting trust, for example; it's actually quite easy to do.

In conclusion, it's important to note that language is not a set of labels fixed to the world, but serves multiple purposes; even as a description of the world, we rarely need a 'full' description, but pick a description to suit various purposes. And there will be notions of value included in many words which are not straightforwardly 'value' terms. Moreover, there will be different implications, and different connotations, for different people. Hence, in looking for definitions of key terms we may not need to get 'the' definition. Rather, it may be better to flag up possible misunderstandings, and make sure that common language does not elide complexity and mask disagreement. The masking of disagreement may occur where codes of ethics are trying to formalise language. Glossaries can be helpful, but not if they shoehorn complex concepts into a box; and it would often be useful to note the difficulties of producing a simple, standard definition.

Chapter 8
Some Suggestions for How to Proceed

Abstract This final chapter makes some provisional suggestions for the development of codes of ethics based upon the discussion so far. This will be of necessity
incomplete, but there is a need to contribute to ongoing debate. Any code of ethics
needs to be embedded well into an organisation and its culture, and specific ways in
which codes of ethics for AI might face problems are indicated. Procedures for
drawing up and implementing codes need to take note of diversity of thinking style
and of experience in participants. The problems of transparency inherent in the
operation of some AI, together with the important public concerns about the impact
of AI, means that maximising transparency and openness in codes of ethics,
appropriate to a particular organisation, is highly desirable. Codes of ethics need
to balance attention to abstract principles with specificity, especially in AI where
application of ethical ideals must be translatable into concrete practice. Procedures
for revision and critique of codes are essential. Ethical discussion leading up to
codes of ethics, as well as the codes of ethics themselves, must include consideration of issues concerning boundaries of human functioning, which is a key issue in
AI and which may be left out of some ethical debates. Particular attention to the
implications of replacing or extending human agency, and impacts upon complex
social systems, would be useful. Lastly, the Asilomar AI Principles are briefly
discussed, as an example of a recent attempt to produce principles intended to
stimulate debate and discussion about beneficial and ethical AI.

8.1 Organisations and Codes

A code of ethics are only as good as its organisational backing. The way in which
development of codes of ethics for AI is managed, and how such codes are
implemented, will be one element of such organisational integrity, for good or for
ill. These points apply to codes of ethics in general; but some problems are likely to
be especially acute in AI.

Codes of ethics may function more as window dressing than real applied policy.
In AI, where fears abound, the temptation to produce a wonderful sounding code of
ethics simply to ward off criticism may be especially acute. Conspicuous virtue can
also be a trap for the content of the codes: overstating certain values or virtues might

© Springer International Publishing AG 2017   99
P. Boddington, *Towards a Code of Ethics for Artificial Intelligence*,
Artificial Intelligence: Foundations, Theory, and Algorithms,
DOI 10.1007/978-3-319-60648-4_8

make it impossible for the good to combat the bad. The control problem in AI makes this an especially important issue.

Given how AI challenges the basis of standard professional codes of ethics, there is particular reason for hard thinking about how to develop and implement such a code. There should be explicit attention to how values are imbued in practices and how they may be present subliminally in the language and framing around codes and regulations.

Appointing someone with specific responsibility for the institutional memory, to keep track of the organisation's own history and thinking regarding value issues might be valuable and useful in an area of such rapid change as AI, especially given how technological development can lead to incremental changes in value which over time may cross boundaries which were once 'lines in the sand', although this might be unfeasible for small organisations.

We've seen how the control problem in AI affects the authoritative basis of professional power in AI. It would be wise for organisations clearly to state these difficulties, ideally specified in relation to the specific forms of AI that concern them. We've also seen how widely some forms of AI may affect and disrupt society. Again, it would be wise for institutions to show awareness of when issues are touching on wider political, social, and cultural issues that are beyond their capacity to address sufficiently, even though these institutions may have a vital role to play in societal dialogues.

Not all organisations will have the same range of value concerns; private industry has different concerns from governmental organisations, and some organisations have more local, others more global, concerns. Precision and self-awareness in such matters is valuable, and likely to go further to gaining public trust than bland statements of very general value. The task of specifying values in relation to concrete particulars should also be easier.

8.2   Procedures for Drawing Up and Implementing Codes

Diversity in participation is needed in drawing up, revising, critiquing and implementing codes of ethics. The potentially transformative nature of AI heightens the need for diverse, constructive and creative input. We need diversity of opinion, thinking style, status, interests, experience, and of position in hierarchies; however, beware of falling into the trap of having 'tick box' quotas for 'diversity'. Consideration might be given to ensuring that diverse personality types are represented, to gain a full range of thinking styles. In addition, subject matter experts from outside the realm of AI, such as lawyers, economists, social scientists, public engagement, and others, will be useful. The inclusion of members who have serious interests outside the world of AI may be useful for maintaining an outside perspective. Attention should be given to the leadership of discussions regarding ethics.

Input from those with expertise in areas such as the social impact of technology, and those who stand in diverse relationships to the technology of AI, would be especially welcome. People with expertise in the history of ideas, and understanding of the historical sweep of changes in both technology and in ethics could make valuable contributions, given the disruptive nature of at least some AI.

Transparency: The problems of transparency inherent in some forms of AI mean that gaining maximum transparency elsewhere whereever possible is particularly desirable. Although private corporations may be chiefly answerable only to themselves, their boards of directors, and shareholders, as much transparency as possible about membership and recruitment is desirable, as well as steps to ensure a measure of independence for those with especial responsibility for ethics within an organisation. This must include openness about the operation of any ethics committee or board. This is especially true for those forms of AI which have wide or ubiquitous impact on the lives of millions or even billions.

Good communication with other bodies, and willingness to participate in public discussions and consultations, would be a virtue. This should include discussions about legal change and development since AI concerns questions of agency and the distribution of responsibility, also key concepts to legal systems.

Revision and critique: There must be provision for the revision of codes, and provision for whistleblowing procedures, and as well, good lines of communication to reduce any need for whistleblowing. Thought should be given to procedures for ascertaining the impact of codes.

Timing: Attention needs to be paid to the timing of discussions drawing up codes. There may be some need for swift responses to issues, but in general, where these issues are concerned, careful thinking which takes time is needed.

8.3   The Content of Codes

This section is not intended to be comprehensive, but merely indicates some suggestions based on discussions earlier in the book.

The specificity of codes: There is always a balance between the generality and precision of codes of ethics. In AI, where codes of ethics relate to the development of AI itself, they need to be in a form such that the engineers will be able to translate them into realisable steps. There may be a tension between producing codes of ethics that retain general principles, and that can be embedded in workable practice. This relates to questions of the distribution of responsibility and tasks throughout an organisation. General ethical statements about 'producing benefit for all' and so on, will simply have no impact unless they can be translated into concrete ways of making a positive difference on the ground. Codes may therefore need to be presented at different levels of specification.

Ethical Uncertainty and Rigid Rules: Can Virtue Ethics Come to the Rescue?

One common response to the difficulty of producing future-proofed codes of ethics in areas of rapid development or contextual uncertainty is to refer to virtue ethics (Atkinson 2009). This recognises the importance of equipping researchers and professionals with the ethical skills to make nuanced decisions in context, to provide careful contextualised interpretation of rules, and to judge when rules are no longer appropriate. For example, the Association of Internet Researchers have suggested a strategy of equipping people with *phronesis*, (practical wisdom) drawing on the Aristotelian conception of this (Aristotle 1999; AoIR 2012).

Note that the AoIR suggests an Aristotelian approach to deal with situations where the right ethical path is unclear. Aristotle is frequently quoted as claiming that in any matter of inquiry, one can only hope to produce the degree of precision which that subject area permits (Aristotle 1999). This is sometimes erroneously used to justify vagueness or a range of acceptable, (yet perhaps mutually incompatible) answers. Yet, for Aristotle, making the appropriate ethical decision was understood as getting the appropriate answer, as hitting a target as closely as possible, and he certainly did not intend to allow for ethical pluralism. The call for *phronesis* as a desiderata in codes of ethics for rapidly developing technologies may not in fact provide an answer, so much as indicate the depth of the problem.

Moreover, for Aristotle, crucially, few actually possess *phronesis*. It indicates wisdom achieved over years; on the point of the rarity of true moral wisdom, he was probably correct. The virtues are habits of thought and action—to do the right thing, in the right situation, with the right motivation and thoughts—and note, these habits are acquired within a stable cultural context, by learning from those older and more virtuous, and with the starting assumption that those embarking on the path to virtue already have a good understanding of ethical action, and a strong motivation to live a good life. The application of virtue ethics in a diverse setting of rapid technological development is questionable to say the least.

Note that many Aristotelian virtues would not fit with current values (e.g., he had slaves and women were kept out of public life). In other words, to talk of having a virtue ethic as a framework is to leave wide open what the virtues are. To know who exhibits *phronesis* we have to be able to identify who the good guys are. (It's interesting that a frequent theme of sci fi is the precise difficulty of knowing who's the good guy and who's the bad guy—this is no coincidence.)

There are foundational issues with an Aristotelian account of the virtues, since it is linked intimately to a teleological account of human nature basing the 'good for man' on the 'function' of mankind, which is our unique nature.

(continued)

But not only is such an account far more controversial in the twenty-first century. One feature of AI is the way in which raises questions about humanity's 'uniqueness' or otherwise, and raises questions about what our 'true nature' really is. By presenting us with such destabilising thoughts, by potentially bringing wide ranging changes to society and to how we interact with the world, AI produces precisely the polar opposite of the relatively stable and small world of ancient Athens in which Aristotle could write with confidence about the virtues.

The level of specificity and detail of codes will also be relative to the specific forms of AI in question: self-driving cars for international export, robots for local use in care homes, algorithms for use in search engines, all present different challenges. There may or may not be need to address global or cross cultural issues. Indeed, fine tuning the values of AI may well involve looking very closely at localised values and priorities.

Responsibility: Questions of responsibility and accountability, their distribution within an organisation, and attention to how the implementation of AI itself affects responsibility and accountability, should be included.

AI in context: Attention to issues concerning AI in use is important, although may be difficult where the particular context of application is not specified in advance. It may be important to consider procedures for liaison with others concerning the downstream application of AI and how it might impact upon complex settings.

AI and the law: Attention to legal regimes local and internationally will of course be needed; a lesson that can be learned from elsewhere is to raise the question of whether or not legal loopholes are being used exploitatively.

Support for further research, and active collaborations, would be welcome, including research into the ethical issues, and how best to further constructive developments in the ethics of AI.

8.4 Thinking About Ethical Issues in Developing and Implementing Codes of Ethics

Benefits of AI: It must be explicitly recognised how hard it is to assess the 'benefits' and 'harms' of AI, and how differently these may be understood; given the potentially transformative nature of AI, this especially important.

AI, agency, and idealisation: As described earlier, it would be a good idea to take note of the particular dangers of idealisation of human and machine agency in discussions of the ethics of AI, and of the question of how hype can distort thinking in AI.

Checking for incompleteness of ethical discussion: One way of attending to distortions of thinking is by implementing procedures to consider the different viewpoints of all those affected by particular developments in AI, and different ways that the ethical issues may be understood.

Including consideration of boundary issues in ethics: We've seen how different thinking styles in ethics include or exclude consideration about issues of boundaries, especially relevant in AI concerning boundaries of human agency and action and even physical boundaries. This often relates to debates about what is 'natural' or the issues which may inspire 'disgust'. Although many philosophers may argue against the relevance of these issues, they may be particularly important in AI, and particularly important for some of groups of people whose voices may currently be less heard in academia, as discussion earlier indicated.

Replacing or surpassing humans: Specific attention to the impact of replacing or supplementing human agency with machine agency on humans, and on how this then affects wider social systems, would be useful.

The limits of expertise: This will include specific recognition of how there might be wider impacts beyond the knowledge and immediate control of AI professionals. This will include recognition of the problems within the AI community of combatting unwise or even malicious AI.

Language and communication: As we've seen, there is a need for precision and understanding regarding AI and in particular some key terms such as autonomy and transparency. It's important to bear in mind the different ways such terms may be understood and implemented, and to check and recheck for good communication.

The public: There's a particular need to pay attention to how issues are communicated to members of the public, or rather, the many different publics. It is preferable to think of developing a dialogue with members of the public, rather than simply 'educating' them about AI.

8.5   Asilomar AI Principles

The recently developed Asilomar AI Principles, drawn up in January 2017 (Future of Life Institute), can serve as an example of an initiative to begin to draw up principles for AI, including ethical principles. They were specifically intended to promote discussion, as is appropriate, given the early stage of consideration of value issues in the development of AI, and given the desirability of wide involvement in the debates around the ethics of AI.

8.5.1   The Process of Producing the Principles

The Principles were discussed by participants at a conference in Asilomar organised by the Future of Life Institute. The process of developing the Principles is described on their website. The basis for inclusion in this conference is not specified, but it appears to involve various prominent people working in AI as well as those from other disciplines, including law, philosophy, economics, industry, and social science. Many participants were holders of Beneficial AI grants awarded by the FLI; as invitations were extended to Principle Investigators, I was not myself present. Although there was a range of expertise involved, the participants cannot be said to be 'representative' of their particular areas of specialisation in any formal way, in the absence of a specific process for ensuring representativeness. Principles drawn up by the prominent have their place, but may miss elements that might be uncovered by the inclusion of those with less visible power. Bearing in mind our discussions earlier about diversity and the facilitation of group intelligence, the list of names of attendees suggests that that approximately 20% of participants were women.

Prior to the conference, members of the FLI compiled various recent reports into AI and from these, distilled a list of opinions about how society should best manage AI, from this list they distilled out a set of principles that expressed some level of consensus. These were then sent out to conference participants in an iterative process that saw a revised list of principles put up for discussion at Asilomar, and refined again over several days of debate. Attendees finally voted on each Principle and only those with 90% approval were included in the final set of 23 Principles. The Principles are available online and those who wish to can add their names.

The process thus was designed to achieve consensus; this is of course one method of generating material for discussion, but debate is also especially worthwhile in contested areas, and it would have been interesting to know if there were any issues on which firstly, the reports initially used to draw up the Principles, and secondly, the Asilomar participants, were in serious disagreement. Reports with minority opinions clearly expressed can provide valuable material for debate. Note, too, that consensus may sometimes be achieved at the expense of abstraction and of choosing words which may mask disagreement. The FLI website recognises that the Principles are open to varying interpretations and are likely incomplete, and considers them aspirational.

The 23 Principles are divided into three sections: Research Issues, Ethics and Values; and Longer Term Issues. Below I give brief commentary on aspects of these Principles, drawing on the discussions throughout this book.

8.5.2 Research Issues in the Asilomar Principles

1. Research Goal: The goal of AI research should be to create, not undirected intelligence, but beneficial intelligence.
2. Research Funding: Investment in AI should be accompanied by funding for research on ensuring its beneficial use, including thorny questions in computer science, economics, law, ethics, and social studies (with example questions added).

3. Science-Policy Link: there should be constructive and healthy exchange between AI researchers and policy makers.
4. Research Culture: A culture of cooperation, trust and transparency should be fostered among researchers and developers of AI.
5. Race Avoidance: Teams developing AI systems should actively cooperate to avoid corner-cutting on safety standards.

It is hard to disagree with any of these Principles. But are there ways they could be improved? One major omission in the groups mentioned are members of the public. This is unfortunate, especially given the difficulty of defining the key notion of what would constitute 'benefit' in anything, especially AI, which may drive deep into the heart of our entire account of value and meaning.

Notwithstanding the consensus-driven and aspirational nature of the Principles, some recognition of the institutional, financial and policy burden of these Research Principles would be useful in any development of them. Who will provide the funding for research into the beneficial use of AI? Consider the case of private corporations doing such research. It's common for such corporations to aspire to ethical principles—but they also have duties towards their shareholders and a need to make a profit, or at least keep solvent. Moreover, if research into beneficial uses of AI does come from private sources, this will leave many questions open, given the contested nature of what counts as a benefit. Would a private company be more likely to think that a 'benefit' involves steps which lead the populace to be dependent upon their products and services, or those of their corporate friends? Some indication of what *specific* issues there might be in AI would be welcome too.

And while recognising that there is not space for detail in such Principles, much of what is indicated here will depend upon the institutional and governmental context within which AI is being developed. Principle 4 regards Research Culture, but this requires robust and healthy institutions; this could be mentioned; and a note about why AI in particular has a difficulty with cooperation and transparency would be useful and would help give more precise direction to any thoughts about the implementation or further elaboration of the Principles.

8.5.3   Ethics and Values in the Asilomar Principles

6. Safety: AI systems should be safe and secure throughout their operational lifetime, and verifiably so where applicable and feasible.

Comment: it's hard to argue with this one. There are of course challenges concerning assessing safety with regard to complex human-AI interactions

7. Failure Transparency: If an AI system causes harm, it should be possible to ascertain why.

Comment: As an ideal, this is laudable. But there is uncertainty if it can be achieved technically. There are various moves available to deal with cases where the cause of harm is unverifiable, for example in law with regimes of strict liability, where attributions of the cause of harm are not necessary to assign responsibility for redress. I'd suggest: 'AI systems should be developed so that, as far as possible, it will be possible to ascertain the causes of any resulting harm, and steps taken to assign responsibility and redress where this is not possible. Full consideration to what constitutes harm should be given'.

8. Judicial Transparency: Any involvement with an autonomous system in judicial decision-making should provide a satisfactory explanation auditable by a competent human authority.

Comment: It's pretty much up to judicial systems to decide on this one, and such questions are currently receiving much scrutiny, as we've seen in Wisconsin vs. Loomis. Cooperation between legal scholars and law makers, and the AI community, is of course essential. AI needs to be fully integrated into human systems, and legal systems already have their own set of ideals of operation and notions of procedural justice, which AI must only enhance, not weaken.

9. Responsibility: Designers and builders of advanced AI systems are stakeholders in the moral implications of their use, misuse, and actions, with a responsibility and opportunity to shape those implications.

Comment: again, a laudable sentiment, and aspirational. As I've argued, figuring out how to distribute and maintain responsibilities across a large network of often loosely connected people and institutions is a very vexed question. The notion of responsibility is also rather elastic and has various uses in context; one common reason why people resist calls to responsibility is because of how swiftly it leads, or may be perceived to lead, to blame. There are some good reasons for this: among them, that attribution of responsibility without adequate control is a major dimension of work place stress, with concomitant serious health effects (Marmot et al. 1997). Although it is desirable for designers and builders of AI to consider the misuse of their systems, calls to responsibility might be counterproductive if done in ways which suggest responsibility for problems over which they have scant or no realistic control.

10. Value Alignment: Highly autonomous systems should be designed so that their goals and behaviours can be assured to align with human values throughout their operation.

Comment: this is of course again aspirational. I would add explicit reference to the embedding of autonomous systems within human social and work settings and the necessity of understanding the possible complexities here. As Francesca Rossi stated in an interview on the Principles, '... when you have human and machine tightly working together, you want this to be a real team. So you want the human to be really sure that the AI system works with values aligned to that person. It takes a lot of discussion to understand those values' (Conn 2017c).

There is a tendency in the Principles to talk of AI as a whole. I'd also add that value alignment will be highly specific to each instance and context of use. In any event, it will only be in examining specific circumstances that value alignment can occur. This process could at its best even improve the value alignment for certain activities, if it involves clarity and explicitly operationalising underlying values.

11. Human Values: AI systems should be designed and operated so as to be compatible with ideals of human dignity, rights, freedoms, and cultural diversity.

Comment: Again, naturally human dignity, rights and freedoms should be aspired to. However, the knotty question as always is, how do you achieve dignity, and which rights and freedoms? This vexed problem can often be cut to size by again noting that many AI systems will operate in certain contexts only. The extremely complex question of cultural diversity has been addressed earlier. Respecting people from other cultures is a given; it's part and parcel of a universalist ethic. Yet allowing unfettered cultural diversity of values is, as a matter of verifiable empirical fact, inconsistent with implementing certain understandings of human rights; cultures concern values. A set of Principles for AI can't be expected to sort this one, but given that AI professionals deal all day long with ironing out bugs and inconsistencies in computer programmes, they might have noticed the tensions here.

12. Personal Privacy: People should have the right to access, manage and control the data they generate, given AI systems' power to analyse and utilize that data.

Comment: this is an example of a Principle which does at least mention the relevance of AI in particular to the issue. A major conceptual question is what counts as 'data they generate': for example, since individual data may be pooled, data needs to be analysed with considerable sophistication so it's not necessarily clear what the basis and extent of individual rights are. However, these questions are questions for data analysis in general and attention to the particular role of AI might add clarity. The role of AI can indeed involve helping to address the issues of individual control over data with AI driven solutions.

13. Liberty and Privacy: The application of AI to personal data must not unreasonably curtail people's real or perceived liberty.

Comment: Again, mention of the role of AI in escalating concerns about personal data use, and attention to any specific responsibilities that this produces, would tighten this Principle from a general issue about privacy to one focused on AI. Moreover, liberty is frequently in tension with AI; posing this Principle in the form of raising the question about what counts as 'reasonable' curtailment, and whether AI has anything to do with shifting conceptions of 'reasonable curtailment of liberty' in one direction or another, would be welcome. Everything hangs on what is construed as 'unreasonable'.

14. Shared Benefit: AI technologies should benefit and empower as many people as possible.

Comment: there's no reason given to explain why AI has any particular reason to be concerned with benefit and empowerment in general. If AI is produced by private companies, it will be in their economic interests to ensure good corporate reputation and a consistent customer base who can afford their products, but that they have any further duties to general benefit is unclear. But if AI were responsible for the loss of benefits or power, this does give reason for its producers to guard against this, and mitigate or offer redress. Again, a statement which more explicitly cited ways in which AI might reduce the power or benefits of people, and looked to specific ways of combatting this, might provide a more precise and hence firmer basis for moving forward.

15. Shared Prosperity: The economic prosperity created by AI should be shared broadly, to benefit all of humanity.

Comment: it's left entirely unclear how this could be achieved. I would suggest that some clarity about whose responsibility this is would be welcomed. We might be left in a situation where governments are forced to mop up the economic and social mess created by AI-induced redundancies and escalating wealth disparities.

A set of aspirational Principles without any indication of whose responsibility it might be to bring them about, or how this is to be implemented, is to that extent weaker. It's very early days for AI, but yet, Principles for AI would have more weight, the more they can be linked to concrete specifications.

16. Human Control: Humans should choose how and whether to delegate decisions to AI systems, to accomplish human-chosen objectives.

Comment: the Principle of keeping human control and choice over delegation of decisions is good; but note that it's ambiguous about whether this means 'some human should choose, not a machine' or 'all humans should be able to choose'—the former case might still mean that many other humans are subject to human-machine systems. The reach of AI in certain areas indeed makes this likely. This is an issue for the differential spread of power and influence in society under AI, and mention of this in any Principles would be welcome. This links of course to Principles 14 and 15 which concern the differential benefits of AI. Loss of control over aspects of one's life is one such possible harm of AI for many.

17. Non-subversion: The power conferred by control of highly advanced AI systems should respect and improve, rather than subvert, the social and civic processes on which the health of society depends.

Comment: this is an interesting Principle which raises an abundance of issues. It points to how far AI can reach into our lives. The problem raised suggests that there is a need for a variety of groups overseeing and commenting on how AI is interacting with our social and civic processes—this is important to recognise the importance of debate, and the difference of viewpoints possible here, as well as the impact upon views and levels of influence of issues like funding sources, representation in such groups, and so on.

There are many examples of how developments in AI are likely to impact upon social and civic processes, too many to illustrate here. The recent EU and Whitehouse reports raise concerns about its possible impact upon taxation, and the need for government intervention and support in developing essential areas to support the long term overall social interests of AI, where private financial interests may have insufficient individual reason to invest (European Civil Law Rules in Robotics 2016; Preparing for the Future of Artificial Intelligence 2016). The concerns of the EU with harmonisation (European Civil Law Rules in Robotics 2016) indicate a wish to step in before advances in other jurisdictions force those lagging behind to fit in with others. Hence, international relationships are also implicated. Long term, and global, thinking is needed. Yet, our current civic processes have been noted to work against the need for longer term thinking about AI (Conn 2017d).

18. AI Arms Race: An arms race in lethal autonomous weapons should be avoided.

Comment: achieving this will be challenging. One way to avoid an arms race is to let the enemy win; presumably this is not what those who signed these Principles had in mind. A topic for another book, or indeed, for many volumes.

8.5.4   Longer-Term Issues in AI

There are various longer term issues included in Principles 19–23. Just one will be discussed here.

23. Common Good: superintelligence should only be developed in the services of widely shared ethical ideals, and for the benefit of all humanity, rather than one state or organisation.

Comment: this seems to suggest that ensuring that superintelligence can be produced to align with widely shared ethical ideals is possible. And much hangs on how any such 'widely shared' ideals are identified. Ideals held by large minorities are nonetheless 'widely shared'; ideals held by majorities can do untold damage to minorities.

8.5.5   General Comments on the Asilomar Principles

These Principles are of course an early step in the process of thought about beneficial AI. There are advantages to attempting to achieve consensus, but nonetheless, expressing some of these Principles in terms of the questions to be raised around these points, rather than as statements expressed with a degree of certainty, might help to open up and continue discussion, without forgoing consensus.

Likewise, although aspirational, it would be beneficial to try to focus them as closely as possible on the distinctive or typical role of AI, and to avoid statements of very general principle which raise issues which are not unique to AI, but might apply to any technology, or to any commercial or industrial enterprise. Contrariwise, the Principles tend to refer to AI in general, which then implies we need to consider value issues for AI in general, whereas very often, the value issues we need to consider are much more local and contextualised—and therefore, to that extent, easier to address.

More explicit reference to the way in which AI will be closely embedded in complex human systems, and therefore, to that extent more complex to assess, would be helpful in indicating the necessary direction of much future work. For this and for other reasons indicated above, although the work of professionals in AI is absolutely necessary, including technological work on issues such as safety, verification, and transparency, emphasis also needs to be given to the role of others, including members of the public, and the representativeness of those involved in discussions about the ethics of AI. This is certainly the case given the difficulty, discussed throughout this book, of ascertaining what constitutes 'benefit' in the development of AI; the Principles could usefully indicate awareness of this issue.

References

Adams G, Balfour D (2014) Unmasking administrative evil, 4th edn. Routledge, Abingdon

Admin (2017) SCOTUS asks public defender: does use of COMPAS at sentencing violate due process? On point, Wisconsin State Public Defender, 6 March 2017

Agar N (2004) Liberal eugenics: in defence of human enhancement. Wiley Blackwell, Oxford

Ansar Burney Trust (2013) Almost three thousand under age camel jockeys missing. ansarburney. org. http://ansarburney.org/almost-three-thousand-underage-child-camel-jockeys-missing/. Accessed 20 Jan 2016

Anti-Slavery International (2006) Information on the United Arab Emirates (UAE) Compliance with ILO Convention No. 182 on the Worst Forms of Child Labour (ratified in 2001) Trafficking of children for camel jockeys

AoIR (2012) Ethical decision-making and internet research: recommendations from the AoIR Ethics Working Committee (Version 2.0)

Archard D, Lippert-Rasmussen K (2013) Applied ethics. In: Lafollette H (ed) International encyclopedia of ethics. Blackwell, Abingdon

Aristotle (1999) Nicomachean ethics (trans: Irwin T), 2nd edn. Hackett Publishing, Indianapolis, IN

Aspray W (2016) Women and underrepresented minorities in computing: a historical and social study. Springer, Heidelberg

Atkinson P (2009) Ethics and ethnography. Twenty First Century Soc 4(1):17–30. doi:10.1080/17450140802648439

Augustine (2014) Confessions (trans: Hammond CJ-B). Loeb Classical Library 26. Harvard University Press, Cambridge, MA

Barrett L, Dunbar R, Lycett J (2002) Human evolutionary psychology. Princeton University Press, Princeton, NJ

Bartholomew J (2015) Hating the daily mail is a substitute for doing good. The Spectator, 18 April 2015

Bass D (2017) No big deal, right? We're a meritocracy. Bloomberg Technology, 22 March 2017

Bazerman MH, Tenbrunsel AE (2011) Blind spots: why we fail to do what's right, and what to do about it. Princeton University Press, Princeton, NJ

Bear JB, Woolley AW (2011) The role of gender in team collaboration and performance. Interdiscip Sci Rev 36(2):146–153

Beauchamp TL, Childress JF (2001) Principles of biomedical ethics. Oxford University Press, New York, NY

© Springer International Publishing AG 2017
P. Boddington, *Towards a Code of Ethics for Artificial Intelligence*,
Artificial Intelligence: Foundations, Theory, and Algorithms,
DOI 10.1007/978-3-319-60648-4

BeeBee H, Saul J (2011) Women in philosophy in the UK: a report by the British Philosophical Association and the Society for Women in Philosophy in the UK. British Philosophical Association, Society for Women in Philosophy UK, London. http://bpa.ac.uk/uploads/2011/02/BPA_Report_Women_In_Philosophy.pdf

Belloc H (1907) Cautionary tales for children: designed for the admonition of children between the ages of eight and fourteen years. Harcourt, New York, NY

Benatar SR, Singer PA (2000) A new look at international research ethics. BMJ 321(7264):824

Berg P, Baltimore D, Brenner S, Roblin RO, Singer MF (1975) Summary statement of the Asilomar conference on recombinant DNA molecules. Proc Natl Acad Sci 72(6):1981–1984

Berners-Lee T (2017) Three challenges for the web, according to its inventor. World Wide Web Foundation. http://webfoundation.org/2017/03/web-turns-28-letter/. Accessed 16 Mar 2017

Boddington P (2011) Ethical challenges in genomics research: a guide to understanding ethics in context. Springer, Heidelberg

Boddington P (2012) Data sharing in genomics. In: Boddington P (ed) Ethical challenges in genomics research: a guide to understanding ethics in context. Springer, Berlin, pp 195–216

Boddington P (2016) Big data, small talk: lessons from the ethical practices of interpersonal communication for the management of biomedical big data. In: Mittelstadt DB, Floridi L (eds) The ethics of biomedical big data. Springer, Cham, pp 277–305. doi:10.1007/978-3-319-33525-4_13

Boddington P, Hogben S (2006) Working up policy: the use of specific disease exemplars in formulating general principles governing childhood genetic testing. Health Care Anal 14(1):1–13

Boddington P, Podpadec T (1992) Measuring the quality of life in theory and in practice: a dialogue between philosophical and psychological approaches. Bioethics 6(3):201–217

Boddington P, Räisänen U (2009) Theoretical and practical issues in the definition of health: insights from Aboriginal Australia. J Med Philos 34(1):49–67

Boden M, Bryson J, Caldwell D, Dautenhahn K, Edwards L, Kember S, Newman P, Parry V, Pegman G, Rodden T, Sorell T, Wallis M, Whitby B, Winfield A, Parry V (2011) Principles of robotics. Engineering and Physical Sciences Research Council (ESPRC), Swindon

Boorse C (1975) On the distinction between disease and illness. Philos Public Aff 5:49–68

Bostrom N (2014) Superintelligence: paths, dangers, strategies. Oxford University Press, Oxford

Bowden P, Surma A (2003) Codes of ethics: texts in practice. Prof Ethics 11(1):19–37

Bowie N (2009) Organisational integrity and moral climates. In: Brenkert GG (ed) Oxford handbook of business ethics. Oxford handbooks online. Oxford University Press, Oxford

Bridges J, Nicholson C, Maben J, Pope C, Flatley M, Wilkinson C, Tziggili M (2013) Capacity for care: meta-ethnography of acute care nurses' experiences of the nurse-patient relationship. J Adv Nurs 69(4):760–772

Bridges J, Wilkinson C (2011) Achieving dignity for older people with dementia in hospital. Nurs Stand 25(29):42–47

Brook P (2015) The DIY robots that ride camels and fight for human rights. Wired, 03 March 2015

Brynjolfsson E, McAfee A (2014) The second machine age: work, progress, and prosperity in a time of brilliant technologies. W.N. Norton, New York, NY

Brynjolfsson E, McAfee A, Spence M (2014) Labor, capital, and ideas in the power law economy. Foreign Aff 93(4):44

Bryson J (2012) The making of the EPSRC principles of robotics. AISB Q 133(Spring 2012):14–15

Butler J (1827) Fifteen sermons preached at the Rolls Chapel. Hilliard, Grey, Little and Wilkins, Boston, MA

Cairo Declaration on Human Rights in Islam (1990) Aug. 5, 1990, U.N. GAOR, World Conf. on Hum. Rts., 4th Sess., Agenda Item 5, U.N. Doc. A/CONF.157/PC/62/Add.18 (1993), vol Aug. 5, 1990, U.N. GAOR, World Conf. on Hum. Rts., 4th Sess., Agenda Item 5, U.N. Doc. A/CONF.157/PC/62/Add.18 (1993). University of Minnesota Human Rights Library

Campbell R (2015) Moral epistemology. The Stanford encyclopedia of philosophy. Stanford University, Stanford. http://plato.stanford.edu/archives/win2015/entries/moral-epistemology/

Castilla EJ, Benard S (2010) The paradox of meritocracy in organizations. Adm Sci Q 55 (4):543–676. doi:10.2189/asqu.2010.55.4.543

Caulfield T, Condit C (2012) Science and the sources of hype. Public Health Genomics 15 (3–4):209–217

Caulfield T, McGuire AL, Cho M, Buchanan JA, Burgess MM, Danilczyk U, Diaz CM, Fryer-Edwards K, Green SK, Hodosh MA (2008) Research ethics recommendations for whole-genome research: consensus statement. PLoS Biol 6(3):e73

Chambers T (1999) The fiction of bioethics (Reflective bioethics). Routledge, New York, NY

Christakis N, Fowler J (2010) Connected: the amazing power of social networks and how they shape our lives. Harper Press, London

Clark J (2016) Artificial intelligence has a sea of dudes problem. Bloomberg Technology, 23 June 2016

Clarke S (2005) Future technologies, dystopic futures and the precautionary principle. Ethics Inf Technol 7(3):121–126. doi:10.1007/s10676-006-0007-1

Cohen JR, Pant LW, Sharp DJ (1992) Cultural and socioeconomic constraints on international codes of ethics: lessons from accounting. J Bus Ethics 11(9):687–700

Conn A (2017a) Gurduth Banavar interview. Future of Life Institute. https://futureoflife.org/2017/01/18/guruduth-banavar-interview/ . Accessed 30 Mar 2017

Conn A (2017b) Anca Dragan Interview. Future of Life Institute. https://futureoflife.org/2017/01/18/anca-dragan-interview/. Accessed 30 Mar 2017

Conn A (2017c) Francesca Rossi interview. Future of Life Institute. https://futureoflife.org/2017/01/26/francesca-rossi-interview/. Accessed 30 Mar 2017

Conn A (2017d) Kay Firth-Butterfield interview. Future of Life Institute. https://futureoflife.org/2017/01/26/kay-firth-butterfield-interview/. Accessed 30 Mar 2017

Crawford R (1980) Healthism and the medicalization of everyday life. Int J Health Serv 10 (3):365–388

Crawford K, Calo R (2016) There is a blind spot in AI research. Nature 538:311–313

Dawson A (2010) The future of bioethics: three dogmas and a cup of hemlock. Bioethics 24 (5):218–225

De Angelis C, Drazen JM, Frizelle FA, Haug C, Hoey J, Horton R, Kotzin S, Laine C, Marusic A, Overbeke AJP (2004) Clinical trial registration: a statement from the International Committee of Medical Journal Editors. Ann Inter Med, 21 Sept 2004

Department of Health (2001) The expert patient: a new approach to chronic disease management in the 21st century. Stationary Office, London

Dingwall R (2008) The ethical case against ethical regulation in humanities and social science research. Twenty First Century Soc 3(1):1–12. doi:10.1080/17450140701749189

Directorate General for Internal Policies, European Union (2016) European civil law rules in robotics. European Parliament.

Duguid MM, Thomas-Hunt MC (2015) Condoning stereotyping? How awareness of stereotyping prevalence impacts expression of stereotypes. J Appl Psychol 100(2):343

Edmonds D, Warburton N (2010) Philosophy bites. Oxford University Press, Oxford

Executive Office of the President (2016) Artificial intelligence, automation, and the economy. The White House, Washington, DC

Fessler L (2017) We tested bots like Siri and Alexa to see who would stand up to harassment. Quartz, 22 February 2017

Fischer F, Forrester J (1993) Editor's introduction. In: Fischer F, Forrester J (eds) The argumentative turn in policy and planning. Duke University Press, Durham, NC, pp 1–17

Fitzpatrick M (2001) The tyranny of health: doctors and the regulation of lifestyle. Routledge, Abingdon

Floridi L (2013) Distributed morality in an information society. Sci Eng Ethics 19(3):727–743

Floridi L, Sanders JW (2004) On the morality of artificial agents. Minds Mach 14(3):349–379

Ford KM, Hayes PJ, Glymour C, Allen J (2015) Cognitive orthoses: toward human-centered AI. AI Mag 36(4):5–8

Freud S (2002) Civilisation and its discontents (trans: McLintock D). Penguin Classics. Penguin, London

Frey C, Osborne M (2013) The future of employment: how susceptible are jobs to computerisation? Oxford Martin School, University of Oxford, Oxford

Frith CD (2012) The role of metacognition in human social interactions. Philos Trans R Soc B 367 (1599):2213–2223

Future of Life Institute (2017) Asilomar AI principles. https://futureoflife.org/ai-principles/. Accessed 12 Mar 2017

Glaeser EL (2014) Secular joblessness. In: Teulings C, Baldwin R (eds) Secular stagnation: facts, causes, and cures. Centre for Economic Policy Research (CEPR), London, pp 69–82

Gluckman R (1992) Death in Dubai. http://www.gluckman.com/camelracing.html

Goldman A, Blanchard T (2015) Social epistemology the Stanford encyclopedia of philosophy. Summer 2016 Edition. http://plato.stanford.edu/archives/sum2016/entries/epistemology-social/

Grace K AI impacts. http://aiimpacts.org/. Accessed 12 Mar 2017

Graham J, Haidt J, Nosek BA (2009) Liberals and conservatives rely on different sets of moral foundations. J Per Soc Psychol 96(5):1029

Gulhati CM (2005) A new colonialism? Conducting clinical trials in India. N Engl J Med 352 (16):1633

Haidt J (2013) The righteous mind: why good people are divided by politics and religion. Penguin, London

Hajer M (1993) Discourse coalitions and the institutionalisation of practice: the case of acid rain in Britain. In: Fischer F, Forrester J (eds) The argumentative turn in policy analysis and planning. Duke University Press, Durham, NC

Hammersley M (2009) Against the ethicists: on the evils of ethical regulation. Int J Soc Res Methodol 12(3):211–225

HapMap Iinternational Committee (2004) Integrating ethics and science in the International HapMap Project. Nat Rev Genet 5(6):467

Hassan SR (2017) Pakistan says Facebook vows to tackle concern over blasphemous content. Reuters. https://www.yahoo.com/news/pakistan-says-facebook-vows-to-tackle-concerns-over-blasphemous-170447075--finance.html. Accessed 30 Mar 2017

Hatemi PK, McDermott R (2016) Give me attitudes. Annu Rev Pol Sci 19:331–350

Hawking S, Russell S, Tegmark M, Wilczek F (2014) Stephen Hawking: 'Transcendence looks at the implications of artificial intelligence – but are we taking AI seriously enough?'. The Independent, 1 May 2014

Herxheimer A (2003) Relationships between the pharmaceutical industry and patients' organisations. BMJ 326(7400):1208

Heterodox Academy (2016) http://heterodoxacademy.org/. Accessed 30 Mar 2017

Hetschko C, Knabe A, Schöb R (2014) Changing identity: retiring from unemployment. Econ J 124(5). doi:10.1111/ecoj.12046

Hogben S (2009) It's not easy being green: unpacking visual rhetoric and environmental claims in car, energy and utility advertisements in the UK (2007–08). Lang Ecol 3(1)

Holm S (2004) Like a frog in boiling water: the public, the HFEA and sex selection. Health Care Anal 12(1):27–39

Hong L, Page SE (2004) Groups of diverse problem solvers can outperform groups of high-ability problem solvers. Proc Natl Acad Sci USA 101(46):16385–16389

Houston S (1989) National Aboriginal Health Strategy Working Party. Aborig Isl Health Worker J 13(4):7

IEEE Spectrum (2011) Special report: the singularity. http://spectrum.ieee.org/static/singularity. Accessed 16 Mar 2017

IIIM (2015) Ethics policy for peaceful R&D. Icelandic Institute for Intelligent Machines, Reykjavik, Iceland

Illich I (1976) Limits to medicine: medical nemesis the appropriation of health. Marion Boyars, London

Jackson MO (2010) Social and economic networks. Princeton University Press, Princeton, NJ

Jones M (1999) Informed consent and other fairy stories. Med Law Rev 7(2):103–134. doi:10.1093/medlaw/7.2.103

Joy B (2000) Why the future doesn't need us. Wired 8:238–263

Kant I (1972) The moral law, translation of groundwork for the metaphysics of morals (trans: Paton HJ). Hutchinson, London

Kant I (1998) Religion within the boundaries of mere reason: and other writings. Cambridge University Press, Cambridge

Koepsell D (2009) On genies and bottles: scientists' moral responsibility and dangerous technology R&D. Sci Eng Ethics 16(1):119–133. doi:10.1007/s11948-009-9158-x

Kurzweil R (2001) The law of accelerating returns. http://www.kurzweilai.net/the-law-of-accelerating-returns. Accessed 16 Mar 2017

Kuschke B (2012) Association Belge des Consommateurs Test-Achats ASBL, Vann van Vugt, Charles Basselier v Conseil des ministres Case C-236/09 ECJ. De Jure 45(3):624–630

Laurie GT (2001) Challenging medical-legal norms: the role of autonomy, confidentiality, and privacy in protecting individual and familial group rights in genetic information. J Leg Med 22 (1):1–54

Legal Parenthood (2008) HFEA. http://www.hfea.gov.uk/399.html. Accessed 30 Mar 2017

Levy S (2015) How Elon Musk and Y Combinator plan to stop computers from taking over. Backchannel, 11 December 2015

Lillie M (2013) Camel jockeys in the UAE. Human Trafficking Search. http://humantraffickingsearch.net/wp/camel-jockeys-in-the-uae/

Lippman A (1991) Prenatal genetic testing and screening: constructing needs and reinforcing inequities. Am J Law Med 17:15

Lloyd JV, Schneider J, Scales K, Bailey S, Jones R (2011) Ingroup identity as an obstacle to effective multiprofessional and interprofessional teamwork: findings from an ethnographic study of healthcare assistants in dementia care. J Interprof Care 25(5):345–351. doi:10.3109/13561820.2011.567381

Luciano M, Wainwright MA, Wright MJ, Martin NG (2006) The heritability of conscientiousness facets and their relationship to IQ and academic achievement. Personal Individ Differ 40 (6):1189–1199

MacIntyre A (1984) Does applied ethics rest on a mistake? Monist 67(4):498–513

Mackenzie D, Wajcman J (1999) The social shaping of technology. Open University Press, Milton Keynes

Mackie JL (1977) Ethics: inventing right and wrong. Penguin, London

MacNaughton D (1988) Moral vision. Blackwell, Oxford

Majone G (1989) Evidence, argument and persuasion in the policy process. Yale University Press, New Haven, CT

Manyika J, Chui M, Bughin J, Dobbs R, Bisson P, Marrs A (2013) Disruptive technologies: advances that will transform life, business, and the global economy. McKinsey Global Institute, Washington, DC

Marmot MG, Bosma H, Hemingway H, Brunner E, Stansfeld S (1997) Contribution of job control and other risk factors to social variations in coronary heart disease incidence. Lancet 350 (9073):235–239

Marx K (1858) Fragment on machines. The Grundrisse, pp 690–712

McKerrow RE (1993) Critical rhetoric and the possibility of the subject. In: Angus I, Langsdor L (eds) The critical turn: rhetoric and philosophy in postmodern discourse. Southern Illinois University Press, Carbondale, IL, pp 51–67

McLean B, Elkind P (2013, 2004) The smartest guys in the room: the amazing rise and scandalous fall of Enron. Penguin, London

Mesmer-Magnus JR, Viswesvaran C (2005) Whistleblowing in organizations: an examination of correlates of whistleblowing intentions, actions, and retaliation. J Bus Ethics 62(3):277–297

Milgram S (1974) Obedience to authority. Harper Collins, New York, NY

Mill JS (1863) Utilitarianism. Parker, Son and Bourn, London

Mokyr J (2014) Secular stagnation? Not in your life. In: Teulings C, Baldwin R (eds) Secular stagnation: facts, causes and cures. Centre for Economic Policy Research (CEPR), London

Moore FD (1989) The desperate case: CARE (costs, applicability, research, ethics). JAMA 261 (10):1483–1484

Morgan E (2011) The descent of woman. Souvenir Press, London

Morris W (1893) Useful work versus useless toil. Hammersmith Socialist Society, London

Moutafi J, Furnham A, Paltiel L (2004) Why is conscientiousness negatively correlated with intelligence? Personal Individ Differ 37(5):1013–1022

Muchnik L, Aral S, Taylor SJ (2013) Social influence bias: a randomized experiment. Science 341 (6146):647–651

Müller VC (2012) Introduction: philosophy and theory of artificial intelligence. Mind Mach 22 (2):67–69

Mundy L (2017) Why is Silicon Valley so awful to women? The Atlantic, April 2017

Nevicka B, Ten Velden FS, De Hoogh AH, Van Vianen AE (2011) Reality at odds with perceptions narcissistic leaders and group performance. Psychol Sci 22(10):1259–1264. doi:10.1177/0956797611417259

Newton L (2001) The ethical dilemmas in the biotechnology industry. In: Bowie N (ed) The Blackwell guide to business ethics. Blackwells, Abingdon

Nilsson NJ (1984) Artificial intelligence, employment, and income. AI Mag 5(2):5

Nissenbaum H (2004) Privacy as contextual integrity. Wash Law Rev 79(1):119–158

Nissenbaum H (2010) Privacy in context: technology, policy and the integrity of social life. Stanford University Press, Palo Alto, CA

Novas C, Rose N (2000) Genetic risk and the birth of the somatic individual. Econ Soc 29 (4):485–513. doi:10.1080/03085140050174750

O'Doherty K, Einseidel E (2013) Public engagement and emerging technologies. UBC Press, Vancouver

Oliver J (2003) Charles 'Grey Goo' threat to world. Mail on Sunday, 27 April 2003

OpenAI. https://openai.com/blog/. Accessed 12 Mar 2017

Oye K, Baird L, Chia A, Hocking S, Hutt P, Lee D, Norwalk L, Salvatore V (2013) Legal foundations of adaptive licensing. Clin Pharmacol Ther 94:309–311

Padela AI, Malik AY, Curlin F, De Vries R (2015) [Re]considering respect for persons in a globalizing world. Dev World Bioeth 15(2):98–106

Pagallo U (2011) Killers, fridges, and slaves: a legal journey in robotics. AI Soc 26(4):347–354

Parens E, Johnston J, Moses J (2009) Ethical issues in synthetic biology: an overview of the debates. The synthetic biology project. The Hastings Centre, Garrison, NY

Pariser E (2011) The filter bubble. Viking Penguin, London

Parkin S (2017) The tiny robots revolutionising eye surgery. MIT Technol Rev, 19 January 2017

Peachey P (2010) UAE defies ban on child camel jockeys. Independent. 10 March 2010

Pejman P (2005) Mideast: rehabilitation for retired child camel jockeys gets top priority. http://www.ipsnews.net/2005/05/mideast-rehabilitation-for-retired-child-camel-jockeys-gets-top-priority/

Peterson JB, Flanders JL (2002) Complexity management theory: motivation for ideological rigidity and social conflict. Cortex 38(3):429–458

Piketty T, Goldhammer A, Ganser L (2014) Capital in the twenty-first century. Harvard University Press, Boston, MA

Plato (1974) The Republic (trans: Lee D). Penguin, London

Plows A, Boddington P (2006) Troubles with biocitizenship? Genomics Soc Policy 2(3):1–21

Preparing for the Future of Artificial Intelligence (2016) Washington, DC. https://obamawhitehouse.
    archives.gov/sites/default/files/whitehouse_files/microsites/ostp/NSTC/preparing_for_the_future_
    of_ai.pdf

Rasnai A (2013) Dubai's camel races embrace robot jockeys. The Daily Beast

Rawls J (2009) A theory of justice. Harvard University Press, Cambridge, MA

Reader S (2005) The philosophy of need, vol 57. Cambridge University Press, Cambridge

Rein M, Schon D (1991) Frame-reflective discourse. In: Wagner P, Weiss CH, Wittrock B,
    Wollmann H (eds) Social sciences and modern states. Cambridge University Press, Cam-
    bridge, pp 262–289

Reverby SM (2012) Tuskegee's truths: rethinking the Tuskegee syphilis study. UNC Press, Chapel
    Hill, NC

Rhodes R (1998) Genetic links, family ties, and social bonds: rights and responsibilities in the face
    of genetic knowledge. J Med Philos 23(1):10–30

Robotics and Artificial Intelligence (2016) Science and technology committee. House of Com-
    mons, London

Roff HM (2013) Responsibility, liability, and lethal autonomous robots. In: Routledge handbook
    of ethics and war: just war theory in the 21st century. Routledge, London, pp 352–364

Roff HM (2014) The strategic robot problem: lethal autonomous weapons in war. J Mil Ethics 13
    (3):211–227

Ronay R, Greenaway K, Anicich EM, Galinsky AD (2012) The path to glory is paved with
    hierarchy when hierarchical differentiation increases group effectiveness. Psychol Sci 23
    (6):669–677

Ross T (1988) Super Duper Jezebel. Anderson Press, London

Rotman D (2014) Technology and inequality. MIT Technol Rev, 21 October 2014

Ruskin J (1904) Work. In: The crown of wild olive: four lectures on industry and war. George
    Allen, London

Russell S, Dewey D, Tegmark M (2015) Research priorities for robust and beneficial artificial
    intelligence. AI Mag 36(4):105–114

Scheffler S (1988) Consequentialism and its critics. Oxford University Press, Oxford

Schmitt DP, Long AE, McPhearson A, O'Brien K, Remmert B, Shah SH (2016) Personality and
    gender differences in global perspective. Int J Psychol. doi:10.1002/ijop.12265

Schmundt H (2005) Camel races: robotic jockeys revolutionize desert races. Speigel Online
    International, 18 July 2005 edn

Schnall S, Haidt J, Clore GL, Jordan AH (2008) Disgust as embodied moral judgment. Pers Soc
    Psychol Bull 34(8):1096–1109

Schwitzgebel E (2009) Do ethicists steal more books? Philos Psychol 22(6):711–725

Shaheen S (2017) Meet the camel-riding robot jockeys in Dubai. http://www.vogue.com/
    slideshow/13254503/camel-racing-robot-jockeys-dubai/#19. Accessed 30 Mar 2017

Shanahan M (2015) The technological singularity. MIT Press, Cambridge, MA

Shuster E (1997) Fifty years later: the significance of the Nuremberg code. N Engl J Med 337
    (20):1436–1440. doi:10.1056/NEJM199711133372006

Siddique O, Hayat Z (2008) Unholy speech and holy laws: blasphemy laws in Pakistan –
    controversial origins, design defects and free speech implications. Minn J Int Law 17:303

Smaldino PE (2014) Group-level traits emerge. Behav Brain Sci 37(03):281–295

Snyder K (2016) Language in your job post predicts the gender of your hire. https://
    textio.ai/gendered-language-in-your-job-post-predicts-the-gender-of-the-person-youll-hire-
    cd150452407d#.gz88w5ovr

State of Wisonsin v Eric Loomis (2016) 2015AP157-CR. Wisconsin

Stevenson L, Haberman DL (1998) Ten theories of human nature. Oxford University Press,
    Oxford

Taylor E, Michael K (2016) Smart toys that are the stuff of nightmares [Editorial]. IEEE Technol
    Soc Mag 35(1):8–10

Tenbrunsel AE, Messick DM (2004) Ethical fading: the role of self-deception in unethical behavior. Soc Justice Res 17(2):223–236

Throgmorton J (1993) Planning as a rhetorical activity. J Am Plann Assoc 59(3):334–346

Trash of the Titans (1998) The Simpsons, 9th series, 22nd episode edn

Twigg J (2000) Carework as a form of bodywork. Ageing Soc 20(04):389–411

Vardi M (2012) Artificial intelligence: past and future. Commun ACM 55(1):5

Vinge V (1993) The coming technological singularity: how to survive in the post-human era. In: Proceedings of a Symposium Vision-21: Interdisciplinary Science & Engineering in the Era of CyberSpace, held at NASA Lewis Research Center (NASA Conference Publication CP-10129)—1993

Vollmann J, Winau R (1996) Informed consent in human experimentation before the Nuremberg code. BMJ 313(7070):1445

von Goethe JW (1878) The sorcerer's apprentice. The permanent Goethe. The Dial Press, New York, NY, p 349

Wajcman J (2008) Life in the fast lane? Towards a sociology of technology and time. Br J Sociol 59(1):59–77

Wallich P (2008) Who's who in the singularity: a guide to the singularity true believers, atheists and agnostics. http://spectrum.ieee.org/computing/hardware/who-is-who-in-the-singularity

Warnock GJ (1971) The object of morality. Methuen, London

West RF, Meserve RJ, Stanovitch KE (2012) Cognitive sophistication does not attenuate the bias blind spot. J Pers Soc Psychol 103(3):506

Wilkinson RG, Marmot M (2003) Social determinants of health: the solid facts. World Health Organization, Geneva

Williams BAO (1976) Morality: an introduction to ethics. Cambridge University Press, Cambridge

Williams B (1985) Ethics and the limits of philosophy. Fontana, London

Williams P, Wallace D (1989) Unit 731 the Japanese Army's secret of secrets. Hodder and Stoughton, London

Wiseman H (2016) The myth of the moral brain: the limits of moral enhancement. MIT Press, Boston, MA

Wittgenstein L (1973) Philosophical investigations (trans: Anscombe GEM). Blackwells, Oxford

Woolley AW, Chabris CF, Pentland A, Hashmi N, Malone TW (2010) Evidence for a collective intelligence factor in the performance of human groups. Science 330(6004):686–688

Wright ND, Bahrami B, Johnson E, Di Malta G, Rees G, Frith CD, Dolan RJ (2012) Testosterone disrupts human collaboration by increasing egocentric choices. Proc R Soc Lond B Biol Sci. rspb20112523

Zauzmer J (2017) A scientist's new theory: religion was a key to human evolution. Washington Post, 27 Feb https://www.washingtonpost.com/news/acts-of-faith/wp/2017/02/27/a-scientists-new-theory-religion-was-key-to-humans-social-evolution/

Zimbardo P (2008) The Lucifer effect: understanding how good people turn evil. Random House, New York, NY

Zimbardo P, Boyd J (2009) The time paradox: the new psychology of time. Free Press, New York, NY

Index

A

Adaptive licensing, 69
Administrative evil, 56
Agency, moral, 11, 24, 68, 89, 90
Algorithmic bias, 16
Altman, S., 17, 18
Anca Dragan, 86
Aristotle, 12, 35, 87, 102, 103
Arms race, 55
Asilomar AI principles, 4, 104–111
Asilomar Conference Recombinant DNA
        Molecules, 51
Association of Internet Research (AoIR),
        70, 102
Australian Aboriginal health definition, 46
Autonomous agents, 53
Autonomous machine, 62, 63
Autonomous missile, 30
Autonomous robots, 53, 63
Autonomous systems, 3, 23, 44, 56, 63, 64, 107
Autonomous vehicles, 2, 25, 59, 88, 89
Autonomous weapons, 40, 44, 55, 110
Autonomy, 21, 24, 28, 30, 44–48, 65, 81, 95,
        96, 104

B

Banavar, G., 33
Berners-Lee, T., 60, 76
Bias, 3, 10, 15–17, 23, 43, 54, 78, 80, 96
Biotechnology, 32, 61
Boiling frogs, 57

C

Cairo Declaration of Human Rights, 77
Camel jockeys, 30, 31
Camel jockeys, robot, 94
Charter of Fundamental Rights of the European
        Union, 77
Code of professional ethics, 41, 42, 47, 59
Commercial organisations, 40
Communications, 22, 62, 78, 79, 81, 92–95,
        101, 104
COMPAS, 26
Conscientiousness, 25, 82
Consensus, 57, 62, 64, 105, 106, 110, 111
Consequentialism, 68, 90
Consequentialist theories, 8
Control problem, 29, 53, 63, 89, 94, 99, 100
Crime, 76
Cultural diversity, 108
Cultural relativism, 23

D

Data management, 29
Data privacy, 35
Data protection, 25
Data sets, 16
Data sharing, 36
Definition of health from Aboriginal Australia,
        46
Deontological theory, 8
Developing technologies, 27–29, 57, 67, 102
Dignity, 21, 46, 77, 108

© Springer International Publishing AG 2017
P. Boddington, *Towards a Code of Ethics for Artificial Intelligence*,
Artificial Intelligence: Foundations, Theory, and Algorithms,
DOI 10.1007/978-3-319-60648-4

Printed in the United States
By Bookmasters